Transitions Between Contexts of Mathematical Practices

Mathematics Education Library

VOLUME 27

The titles published in this series are listed at the end of this volume.

Transitions Between Contexts of Mathematical Practices

Edited by

Guida de Abreu

Department of Psychology, University of Luton, U.K.

Alan J. Bishop

Faculty of Education, Monash University, Melbourne, Australia

and

Norma C. Presmeg

Department of Mathematics, Illinois State University,
Normal, Illinois, U.S.A.

Kluwer Academic Publishers

Dordrecht / Boston / London

A C.I.P. Catalogue record for this book is available from the Library of Congress

ISBN 0–7923–7185–2

Published by Kluwer Academic Publishers,
P.O. Box 17, 3300 AA Dordrecht, The Netherlands

Sold and distributed in North, Central and South America
by Kluwer Academic Publishers,
101 Philip Drive, Norwell, MA 02061, U.S.A.

In all other countries, sold and distributed
by Kluwer Academic Publishers Distribution Centre,
P.O. Box 322, 3300 AH Dordrecht, The Netherlands

Printed on acid-free paper

Printed and bound in Great Britain by MPG Books Limited, Bodmin, Cornwall

CONTENTS

ACKNOWLEDGEMENTS

This book represents a shared journey undertaken by researchers working in different contexts and different situations but sharing similar educational concerns and research aspirations. The course of the journey involved the collaboration of many people and institutions to whom we would like to express our gratitude.

We wish to thank first of all the people who have participated in our research projects, without their contribution this volume would not be possible. Teachers and school managers have allowed us into their classrooms. Parents have allowed us into their homes. Children have been patient in answering our questions and in tolerating our observations. Workers have allowed our presence, even intrusion, into their everyday practices and have given us insights on what they do and why. We are very grateful to all of them for opening their doors and allowing us in.

All the empirical work reported in the book was funded by National Agencies and we also wish to express our thanks to these organisations. Namely:

Catalan Ministry of Education, Fundació Propedagògic, Catalonia – Spain, funded the research presented by Núria Gorgorió, Núria Planas and Xavier Vilella in chapter 2;

Australian Research Council – funded the research reported by Alan Bishop in chapter 3, in a collaborative project undertaken with Gilah Leder, Chris Brew and Cath Pearn;

Fundação Ciência e Tecnologia, Portugal (Grant PRAXIS-PCSH-C-CED-146-96), funded the research reported by Madalena Santos and João Filipe Matos in chapter 4;

ESRC – Economic and Social Research Council in the UK (Grant R000222381), funded the research reported in chapter 5 by Guida de Abreu, Tony Cline and Tatheer Shamsi;

Educational Research and Development Centers Program (PR/ Award Number R306A60001), OERI – U.S. Department of Education, funded the research reported by Marta Civil and Rosi Andrade in chapter 6.

Of course many other institutions, including our own Universities or Schools, have supported the authors' development of the ideas presented in the book by giving us time, space and the resources to undertake the research and to participate in local and international research meetings. We are very grateful for this support and wish to take the opportunity to thank all the colleagues who shared and questioned our ideas.

We also wish to thank those that helped us to transform what we learned in our journeys into a book. Here we would especially like to acknowledge Núria Gorgorió and her colleagues at the Faculty of Education, University Autonoma of Barcelona, for the organisation of the first TIEM 98 (Trimestre Intensiu en Educació Matemàtica) in Barcelona. It provided a rare opportunity for the editors to meet. It was during TIEM, in our offices at the CRM (Centre de Recerca Matemàtica), that the first outline of the book structure and the planning of the first group meeting, which took place during PME in South Africa in 1998, were drafted. We are also in debt to the reviewers of the first outline we submitted to Kluwer. Their comments were certainly provocative, and have helped us to produce the book in the current shape. Thanks to them the group had a most enjoyable meeting pre-CIEAEM 51, in the University College of Chichester in 1999. And, here we cannot omit our gratitude to Afzal Ahmed, the Chair of CIEAEM 51, who provided the infrastructure for our group meeting.

Many other names of colleagues who have supported, guided and challenged us come to our minds. It would make a long list to single each one out. The same will be true about the support each of us has received from our families. To all of them we conclude with a message of deep gratitude and appreciation.

LIST OF CONTRIBUTORS

Guida de Abreu
Department of Psychology
University of Luton
Park Square, Luton, Beds
LU1 3JU
UK

Rosi Andrade
Department of Mathematics
University of Arizona
617 N. Santa Rita
Tucson AZ 85721
USA

Alan Bishop
Faculty of Education
P.O Box 6, Monash University,
Victoria 3800
Australia

Tony Cline
Department of Psychology
University of Luton
Park Square, Luton, Beds
LU1 3JU
UK

Marta Civil
Department of Mathematics
University of Arizona
617 N. Santa Rita
Tucson AZ 85721
USA

Núria Gorgorió
Facultat Ciènciòs de l'Educació
Universitat Autònoma de Barcelona
Edifici G G5-142
Bellaterra
08193 Barcelona
Spain

João Filipe Matos
Departamento de Educação
Faculdade de Ciências
Universidade de Lisboa
Campo Grande, C1
1700 Lisboa, Portugal

Núria Planas
Facultat Ciències de l'Educació
Universitat Autònoma de Barcelona
Edifici G G-5 140
Bellaterra
08193 Barcelona
Spain

Norma Presmeg
Mathematics Department
313 Stevenson Hall
Illinois State University
Normal, IL 61790-4520
USA

Madalena Santos
School: Escola Básica 2-3 de Paço d'Arcos
Research Centre: Centro de Investigação em Educação
Faculdade de Ciências
Universidade de Lisboa
Campo Grande, C1
1700 Lisboa
Portugal

Tatheer Shamsi
Department of Psychology
University of Luton
Park Square, Luton, Beds
LU1 3JU
UK

Xavier Vilella
Facultat Ciències de l'Educació
Universitat Autonoma de Barcelona
Edifici G G5-142
Bella Terra
08193 Barcelona
Spain

EDITORS' PRELUDE

RESEARCHING MATHEMATICS LEARNING: THE NEED FOR A NEW APPROACH

We begin this book by sharing with the reader three vignettes which provide a snapshot of the experiences of learners who have to cope with differences between mathematical practices in their school and out-of-school contexts.

VIGNETTE 1 (BRAZIL, YEAR 5 – PRIMARY SCHOOL)

'I'm the worst, because as I said there's no way I can get it into my head, even though I pay attention' (Abreu, 1993, p. 124).

This was how Severina, daughter of an unschooled sugar-cane farm worker, judged her performance in school mathematics. She entered school at the age of 6. At 14 she was still in year 5. She repeated year 4 three times. After school she worked on the production of manioc flour, and also helped her father in sugar-cane farming during the harvest. She acknowledged that people in sugar-cane farming could do sums: 'Yes, they do, but I think they do sums in their heads like my father. But writing they do not do'. Doing sums orally using out-of-school methods, however, did not have the same importance as using school-written methods, since for Severina, these defined for her the places people can access. Referring to the sugar-cane workers she remarked: 'If they had studied they would not be working in that place. This is an example of those who have never been to school, like my father'. Ironically it is Severina's unschooled father who still helps her with her homework: 'I ask him how much is 3 times 7 or 8 and he answers. How much is 3 plus 12? He answers everything.' The various conflicts – cognitive, affective, valorative – which emerged from the differences seem to remain with Severina.

VIGNETTE 2 (CATALONIA, SPAIN, YEAR 3 – SECONDARY SCHOOL)

'I am wrong in your class. ... I do the same mathematics as boys, but I will not do the same work ... I do not want to be a mechanic. Please can I do mathematics for girls?' (Gorgorió, Planas & Vilella, chapter 2).

Saima, a 16 year-old Indian girl, arrived in Barcelona 9 months before she made the above comment to her classroom teacher. She learned the language quite quickly

G. de Abreu, A.J. Bishop, and N.C. Presmeg (eds.), Transitions Between Contexts of Mathematical Practices, 1–5.

but had more difficulties in understanding the social dynamics. Each day, at the end of the classes, her oldest brother was waiting for her in front of the school. In an interview she explained that her brother had to be sure she had no boy-friends, only girl-friends. If she talked to a boy, or even worse, if a boy talked to her, then she would be obliged to leave school. That was the kind of compromise she had agreed to at home just to succeed in going to school, as her parents at first didn't want her to attend school. In an interview her parents told the teacher they were worried about public schools in Barcelona, because 'they look dangerous in terms of putting too close the girls' culture and the boys' culture'. The father was a doctor and the mother an important member of the Indian community in Barcelona. When she is older, Saima wants to become a teacher 'as you are, Miss Nuria, but only a girls' teacher'. It seems clear that for Saima gender was a very salient dimension in the way mathematical practices were organised. The fact that the mathematical practices of her school did not distinguish between mathematics for boys and for girls exacerbated the difference and the tension between school and home identities.

VIGNETTE 3 (ENGLAND, YEAR 9 – SECONDARY SCHOOL)

It is a Monday morning; a lower set, year 9 group (average age 15 years) are working on their booklets. One of the boys (Mohamed) always appears restless in the class. He talks a lot with his friends and doesn't do much work. He arrived from Bangladesh last year and chats in Bengali to his friends most of the time. His spoken English is rather weak, but he can read some of the text if he wants to. But mostly he guesses what's needed, or copies from his friends. His English writing is poor. As I go to speak to him, he shows me a sheet of paper with algebraic equations on it, all solved correctly. They are at a high year 10 level. He claims to have done them all himself, but says 'I don't understand'. I ask him where he learned to do them. 'In Bangladesh'. So where have these come from? 'My book at home, my father helped me'. I give him a new equation to solve, like the others, and he 'talks through' the solution and solves it correctly, with a beaming smile (Bishop, 1991). The researcher was left with a number of questions about Mohamed's experiences of his school mathematics and the support he receives from this father at home. Why did Mohamed show no interest in his school mathematics? What is it that he claimed not to understand? Why did he bring to the classroom exercises that were not part of the practices in Year 9? Why was Mohamed's father teaching him a kind of mathematics different from the one he was doing at school? Was the father trying to help Mohamed to succeed in the 'English' school? Or, was he trying to help Mohamed to learn what he perceived to be 'proper' mathematics, with the intention of keeping him up with the standard of the home culture?

****

In choosing these vignettes our aims are firstly to introduce some lived experiences of learners which help to justify the need that we address in this book to understand their transitions between contexts of mathematical practices. Secondly, we would like to call attention to the fact that despite the advances of sociocultural theories in the last century, there is still no clear understanding of how to help learners like Severina, Saima, and Mohamed, to become competent participants in school mathematical practices. We believe that the vignettes clearly highlight the fact that the ways they make meanings about their participation in specific mathematical practices involve both cognitive and identity constructions. Ways of participating in the practices seem to have been interpreted as having an impact on who they are. Thus, Severina, viewed written school mathematics as associated with the identity of those who study – the schooled person, who could get proper jobs. Saima viewed certain contents of school mathematics as associated with gender roles and identities. Mohamed did not make explicit the notion of identity, but interestingly, he brought to the school a paper with the mathematics he learnt in his school in Bangladesh. There was a hidden message in his behaviour. While he seemed very disinterested in the current school practices in England (or may be disaffected!), he was still showing interest in school mathematical practices by studying in his old book from the Bangladeshi school. However, issues of identity development have been grossly neglected in sociocultural approaches to learning till very recently.

Also parts of the context of this book are particularly important aspects of globalisation, such as:

- an increase in the relative cultural power of the 'developed' countries and of the multi-national corporations they support,
- an increasing recognition of the diaspora situation, with increasing migration from one educational context into another,
- second or third language learners becoming the norm, with many languages being relegated to the rank of minority language,

These developments are in their turn making new demands on knowledge about mathematics teaching and learning. No longer is it possible to ignore the fact that the majority of the world's learners are learning mathematics in school through their second or third language. No longer can one assume that learning and teaching mathematics happens only in school, nor in a school where everyone comes from the same cultural and social background and speaks the same language. No longer can one assume that everyone everywhere practices the same mathematical knowledge irrespective of their cultural and social background (Barton, 1996; Cobb & Bauersfeld, 1995; D'Ambrosio, 1985; Gerdes, 1996; Lerman, 1994, 1998; Nunes et al, 1993; Zaslavsky, 1995).

New perspectives on mathematical knowledge are developing from the studies on diverse mathematical practices, particularly with marginalised learners, and studies like these form the essential context of this book. In our view these perspectives need to be developed further to meet the democratic ideals of the new mathematics education. Instead of continuing the myth that the only mathematics being learnt by students is that being taught in school mathematics classes, there is now the possibility for learners to develop appropriate and meaningful relationships between the mathematical knowledge they are learning both inside and outside school in a variety of situations. Instead of relying on traditional teaching approaches that have marginalised any other knowledge, culture, and values besides those of formal school mathematics, and therefore those 'other' learners also, there is now the possibility of developing more socially and culturally responsive learning environments, and several projects have started exploring these possibilities (Hollins & Oliver, 1999).

For these possibilities to be realised, much depends of course on changing the formal educational structures that determine and shape the particular mathematics education practice experienced by the students in their schools. That is beyond the scope of this book. However, much also depends upon the development of new and significant conceptualisations. One of the fundamental beliefs that the authors of this book share is that it is the constructs, theories and conceptualisations which sustain and define current mathematics education practices that must be addressed if significant change in those practices is to be achieved.

In particular, the authors have come together because they see the need in mathematics education for richer theoretical perspectives that focus attention on

- mathematics learners in transition and on their practices in different contexts,
- the institutional and sociocultural framing of transition processes, and
- the communication and negotiation of mathematical meanings during transition.

The goals of this book then are twofold. Firstly it aims to offer relevant samples of empirical work which help to identify crucial features and dynamics of the experience of transition in different contexts of mathematical practice. In addition, it aims to offer significant theoretical reflections and accounts of these phenomena from a sociocultural perspective.

Finally, before moving to chapter 1, which presents the book's approach to transitions and outlines the different chapters, it is important to remark that though the challenges Severina, Saima, and Mohamed are facing in their classrooms are not necessarily the same as those faced by all learners, some dimensions of their experience of transition can certainly be similar to those of learners of all ages and in all countries.

REFERENCES

Abreu, G. de (1993). *The relationship between home and school mathematics in a farming community in rural Brazil.* Doctoral dissertation, University of Cambridge, UK.

Barton, B. (1996). Anthropological perspectives on mathematics and mathematics education. In A.J. Bishop, K. Clements, C. Keitel, J. Kilpatrick & C. Laborde (Eds.), *International handbook of mathematics education* (pp. 1035–1053). Dordrecht: Kluwer.

Bishop, A.J. (1991). Teaching mathematics to ethnic minority pupils in secondary schools. In D. Pimm & E. Love (Eds.), *Teaching and learning school mathematics* (pp. 26–43). London: Hodder & Stoughton.

Cobb, P., & Bauersfeld, H. (1995). *The emergence of mathematical meaning*. Hillsdale, New Jersey: Lawrence Erlbaum.

D'Ambrosio, U. (1985). *Sociocultural basis for mathematics education.* Campinas, Brasil: Unicamp.

Gerdes, P. (1996). Ethnomathematics and mathematics education. In A.J. Bishop, K.Clements, C.Keitel, J.Kilpatrick & C. Laborde (Eds.) *International handbook of mathematics education* (pp. 909–943). Dordrecht: Kluwer.

Hollins, E.R. & Oliver, E.I. (1999). *Pathways to success in school: Culturally responsive teaching.* Mahwah, New Jersey: Lawrence Erlbaum.

Lerman, S. (Ed.). (1994). *Cultural perspectives on the mathematics classroom.* Dordrecht: Kluwer.

Lerman, S. (1998). A moment in the zoom of a lens: towards a discursive psychology of mathematics teaching and learning. *Proceedings of the 22nd Conference of the International Group for the Psychology of Mathematics Education, 1*, 66–81

Nunes, T., Schliemann, A., & Carraher, D. (1993). *Street mathematics and school mathematics.* Cambridge: Cambridge University Press.

Zaslavsky, C. (1995). *The multicultural math classroom: bringing in the world.* Portsmouth: Heinemann.

CHAPTER 1

MATHEMATICS LEARNERS IN TRANSITION

GUIDA DE ABREU – *University of Luton*
ALAN BISHOP – *Monash University*
NORMA PRESMEG – *Illinois State University*

1. RECENT SOCIOCULTURAL CONCEPTUALISATIONS AND DEVELOPMENTS

Formal, non-formal and informal mathematics[1] education practices continue to evolve through globalisation and through the use of technology and the WWW. They do so in response to the need for more mathematics to be learnt by increasing numbers of students, both school students and adults. As these practices develop, and as adult education and life-long education grow in importance, along with their mathematical versions, there is an increasing need for mathematics education to move away from ideas and practices based on traditional child development theories and normative ideas. This is particularly important if research in mathematics education is to continue to have relevance and influence in these new and diverse fields of activity.

In the last two decades educational and psychological research studies on social, cultural and political aspects of mathematics learning, have raised awareness of the complexities of the process of learning and using mathematics in specific sociocultural practices (see for instance, Bishop, 1988a, 1988b, 1994; Secada, 1992; Van Oers & Forman, 1998, Cobb & Bauersfeld, 1995; Lerman, 1994). On the other hand such studies have also indicated the potential of this field for informing and developing teaching practices at all levels of mathematics education.

[1] This triad of terms is best defined in Coombs (1985), where formal is what happens in required schooling, non-formal education is what happens in non-required courses and structured educational provision outside and after formal schooling takes place. It could include after-school programs, trade courses, university courses etc. He describes informal education as being non-structured and non-required, such as may be obtained from peers, from TV, libraries, WWW etc. Bishop (1993) applies these definitions to different forms of mathematics education. Nunes et al (1993) distinguish between formal and informal education while Coombs' distinctions separate their 'formal' into his two categories of formal and non-formal. However Nunes et al. (1993) also point out that it is important to bear in mind that informal is defined by exclusion, that is informal mathematics, in their terms, is what is not learned at school. Coombs also makes the additional point that as demands on formal education have increased during the last decade so non-formal and informal education have both expanded to meet the increased need.

G. de Abreu, A.J. Bishop, and N.C. Presmeg (eds.), Transitions Between Contexts of Mathematical Practices, 7–21.

For example, recent studies have:

- documented the wealth of mathematical knowledge accumulated over history by specific cultural groups,
- identified relationships between logical and social organisation of specific cultural tools and individual's thinking processes,
- given some indication of the relationships between the ways knowledge is valued and the mechanisms which can lead to social inclusion or exclusion, and
- informed the development of theories of situated learning.

To achieve such a state of knowledge researchers had to be selective and sometimes have undertaken studies on individual and isolated practices. Some relevant examples here are the studies on 'everyday cognition' focused on out-of-school practices, such as tailoring, farming, cooking, street vending, etc., (see for example the research summarised by Nunes, Schliemann & Carraher, 1993; Barton, 1996).

Focusing on individuals engaged in a particular sociocultural practice has been very important in producing evidence of the existence of legitimate forms of mathematical knowledge other than school mathematical knowledge. However, our concern with these studies is related to the fact that individual learners and societies are not static entities, but are dynamic. Moreover it is our belief that neither ontogenetic (individual development) nor sociogenetic (social group development) aspects of change are properly accounted for by current ethnomathematical or sociocultural theories.

Developmental psychologists working within an individual tradition are also questioning the discrepancy between the explanations produced by researchers and the observed 'facts' in the real world. Siegler (1996) made his point by asking: 'whose children are we talking about'. He doubted it was his children! For him the crucial research challenges lie in explaining these three aspects of learning: **variability**, **choice** and **change**. This is certainly not a problem which is restricted to Siegler's information processing approach. Situated cognition and other sociocultural accounts of cognition also have not provided adequate accounts of any of those three aspects (Abreu, 1995).

Firstly, sociocultural accounts have not yet provided satisfactory accounts of ***variability***. Although a focus on diversity has been a central issue in the agenda of approaches to learning and development from a cultural psychology perspective, like other branches of empirical psychology it has tended to explore differences between groups, and left un-analysed any within-group and within-individual differences. It is unclear why the same person can use mathematics competently in one practice, e.g. street mathematics, and then experience tremendous difficulties in learning the mathematics associated with another practice, e.g. school mathematics. It is also unclear why some people from similar backgrounds show one pattern of performance across practices, e.g. some are competent in both, while some show another pattern, e.g. they succeed only in one. This lack of explanation leaves the situated cognition accounts too vulnerable and opens a space for this diversity to continue being 'explained' in terms of a biological basis.

Secondly, ***choice*** and agency, were not central issues in the theoretical and empirical developments in situated cognition. In general the studies channelled their efforts into demonstrating that individuals and social groups have the ability to learn. The methodological choice for the researchers was a focus on a microanalysis of the mathematical competencies of individuals engaged in specific, very often non-mainstream, and low-status, practices (e.g. street children in Brazil). However, social psychology has for a long time demonstrated that even minority groups are not passive, and it is time for learning theories to try to understand the processes of agency at group and individual level (Moscovici & Paicheler, 1978). As several authors have been stressing, the problems of power, access and transparency of how one becomes a member of a community of practice need to be addressed (see for example Goodnow, 1990; Lave & Wenger, 1991).

Thirdly, sociocultural accounts have been criticised for their limited or biased accounts of ***change***. When focusing on the practices, very often these were described in rather static ways, that is they captured the traditional side of the practice but paid no attention to innovation and change (Abreu, 1998). When focusing on the individual, the tendency is that the accounts of change portray patterns, but less attention is paid to the uniqueness of changes in actual individuals. Also very little attention is paid to any conflicts that may occur between the cultures experienced by the learners inside and outside the school (Bishop, 1994).

Authors following approaches that centre around understanding the emergence of new meanings at the individual level (Cobb, 1995), or at the social group level (Duveen, 1998), suggest that the focus on reproduction of the traditional, i.e. the homogeneous side of cultures and societies, can be linked to particular uses of Vygotsky's ideas. Bruner (1996) also argues that a focus on the cultural symbolic systems is not sufficient to explain learning in modern plural and rapidly changing societies. For him 'nothing is "culture free", but neither are individuals simply mirrors of their culture. (...) Life in culture is, then, an interplay between the versions of the world that people form under its institutional sway and the versions of it that are product of their individual histories' (Bruner, 1996, p.14).

Finally, researchers interested in the emergence of new meanings in mathematics tend to emphasise the importance of communication, negotiation and interpretation (Bishop & Goffree, 1986). Meaning-making processes are however very enigmatic (Wertsch, 1991). Although most authors tend to agree that the meanings that the person brings to a situation influence the course of learning, we still know very little about meanings that are not just cognitive.

Bruner's view that the meanings that a child brings to a situation 'are not to his own advantage unless he can get them shared by others' (1990, p. 13) seems to us extremely important. Contextually-bound and socially shared meanings concerning such phenomena as language use, appropriate behaviour, values, and customs are crucially important factors in learning. They may indeed be more important when intercultural communication and interpretation are involved. For example, Pinxten (1994) characterises what he calls Navajo learning in these terms:

- More emphasis on qualitative ordering and aesthetic aspects and less on quantification and universal statements,
- More stress on orthopraxy (to behave properly, appropriately, and so on) and less on orthodoxy (to share the same contents as the other members of the group),
- More dependence on the persons involved in knowledge transfer, and much less room for a curriculum format and hence for a universal status of knowledge,
- More awareness of the negotiation aspects of each learning situation and less respect for the institutional authority of a teacher. (p. 88)

Walkerdine (1988) in her seminal work 'The Mastery of Reason' also concurs with Bruner in illustrating how Western school practices regulate what comes to be seen as the 'right meaning'. She suggests that schools do not enter into a process of negotiation which helps the learner to construct chains of signification, where concepts and mathematical objects can acquire multiple meanings, legitimated by the contexts in which they are used. Instead, she argues that schooling 'empties' and 'represses' the multiple mathematical meanings acquired outside school in order to replace them with a unique and presumably disembedded meaning.

Evidence from empirical studies, however, suggests that this is not the whole story (see Abreu, 1995; Planas, Vilella, Gorgorió & Fontdevila, 1999; Presmeg, 1998). Learners continue to bring meanings into their mathematics lessons, although most of the time this can occur in 'silence'. Learners clearly negotiate much of the learning process as well as the content being learnt. For example it is also not unusual to hear accounts from well-intentioned teachers about the refusal of their students to use outside school knowledge in the classroom. However what we need to know in order to provide any learner with learning environments conducive to expression, sharing and negotiation of meanings still seems to be an open question.

We believe that:

(1) it is necessary to get insights into the dynamics of mathematics learning of individuals who might behave and apprehend meanings in situated ways, but who certainly move across the different practices and institutions of societies, that are themselves continually in the process of change, and
(2) it is therefore necessary to focus on analyses of how individuals and/or social groups experience their participation in, and transition between, more than one sociocultural mathematical practice.

This then is the focus of this book: the idea of transition in mathematics learning, particularly of mathematics learners in transition, and of their transition between different contexts for mathematics learning and practice.

2. TRANSITION – PRELIMINARY DEFINITION AND WORKING NOTIONS

In a certain way this book also tells a story of a group that has been connected by their shared research interests. Our adoption of the notion of transition as a central construct in our work, and the way it has been evolving, also reflects the developing process of the group. In retrospect most of us feel the way we see transitions now is quite far away from where we started. In this section we tell a bit of this story: how it emerged; where we looked for inspiration; and how we proposed to approach it. We believe that this background is important in enabling readers to evaluate the general validity of the ideas the group has generated.

2.1. THE EMERGENCE OF THE NOTION OF 'TRANSITION' AS AN INTEREST OF THE GROUP

The focus on understanding how participation in home mathematical practices that are distinct from school mathematical practices impacts on the learner has been central to the research of most of the contributors of this book for quite a long time (see for instance Abreu, 1995, 1998; Abreu, Bishop, & Pompeu, 1997; Bishop, 1994; Presmeg, 1988). However, our use of the notion of transition to help to theorise this relationship is recent. Although it is difficult to be precise about when we started using this concept, the re-construction of the history of this book leads us to believe we first used it based on common sense. It seems we took for granted, as many social scientists do, that the meaning(s) of the word was shared during our discussions.

We vaguely defined it in our initial group meeting and outlines. It is certainly not difficult for us to identify the phenomena that for us are disturbing. All the contributors to this book have dedicated a part of their lives to understanding why particular groups of learners have difficulties with their school mathematics learning. Though we manage to provide evidence that these learners are capable of mathematical thinking in their home and other environments, we are still a long way from clarifying for teachers why they continue to have difficulties in more formal learning situations such as at school. This gap in our knowledge is disturbing not only for us as researchers and educators, but also for classroom teachers for example, who reject theories of deficits in the child, only to find themselves in a position of not knowing what to do to help the child progress at school.

The idea that we had in mind was that our problem required investigations that go beyond a focus on single practices. Participation in multiple social practices requires the person to move between them and this movement needs to be understood. From movement between contexts of practices we progressed to the notion of transition. Or indeed 'transitions' in the plural, in the sense that from the beginning we also assumed a person can move between various contexts (e.g. home, school, peer groups, etc.) and also that there exist various types of transitions (e.g. linguistic, social, cultural, etc.).

Other key assumptions in the development of our thinking were firstly, that the movement between practices required the theorisation of both the social environment and the individual learner as dynamic entities. Secondly, in accordance with our view that multiple mathematical practices co-exist in society, we were interested in transitions as bi- or multi-directional trajectories. In taking this view we were departing from a common use of the concept of transition in the traditional developmental psychology stage theories, which meant a replacement of a less-developed, or less-mature, stage and associated ways of thinking, feeling and acting, by a more advanced form.

We take the view, in contrast, that different forms of mathematical knowledge and understanding can co-exist and that replacements when they occur are not necessarily based on a scale of development, but can instead be the result of what particular social groups count as legitimate knowledge. The Western tradition in mathematics education and also Western research in child development has avoided the debate about the legitimacy of knowledge, by assuming the supremacy and superiority of 'science' and 'scientific concepts'. However, the status of science is not unquestionable, neither is it produced in a vacuum. The sociocultural-historical milieu in which it is produced, and the assumptions of those that produce it, do shape the products. Bourdieu (1995) wrote of 'the concrete, complicated ways in which linguistic practices and products are caught up in, and moulded by, the forms of power and inequality which are pervasive features of societies as they actually exist' (pp. 1–2).

Mathematics education curricula do not incorporate all the mathematics knowledge that exists. They contain the 'knowledge that should be taught' and they exclude other forms of mathematics. This division is necessary. There is a limit on what a child or adult can learn at any time, and in any case society's educational institutions have certain specified responsibilities. However, what are highly contestable are both the grounds on which the decisions on inclusion / exclusion are based and also who should be making those decisions. What is also contestable is why the forms of knowledge associated with groups that are less empowered (women, some minority ethnic groups, economically disadvantaged people and nations) tend to be excluded more than others (Apple, 1998).

2.2. RELEVANT CONCEPTIONS OF TRANSITIONS FOR THIS BOOK

Now that we have outlined our group's initial thoughts on transition let us look back at the meanings of the word firstly in **Western Dictionaries**, and then in **psychological** and **sociological / educational** theory. The word transition has a Latin root *transitio*, and according to the Webster's Revised Unabridged Dictionary (1913) it has the following meanings: 'Passage from one place or state to another; change; as, the transition of the weather from hot to cold', 'A direct or indirect passing from one key to another; a modulation' (Music); 'A passing from one subject to another' (Rhetoric); 'Change from one form to another. {Transition rocks}' (Geology). Apart from the meaning in geology the first three meanings attributed to the word did not necessarily imply one-way direction. This also applies to definitions in dictionaries in other European languages (e.g. Portuguese) which put the emphasis of the definition on the passage, though acknowledging that some of these routes are supposed to operate in one direction. A passage that links place a to b can work in both directions unless traffic rules forbid movement in the counter-flow.

The constant in all the definitions is that at least two modalities of a phenomenon exist either contemporarily (places, subjects) or in time-sequences (music, states of weather) and that circumstances require a type of transit and passage between the two. Another interesting feature of the definitions is that they offer examples of transition in subjects where there have been attempts to provide very detailed accounts of the process. Both in music and rhetoric, the passing can occur without individuals being able to describe in any explicit way the phenomenon. For instance, it is not difficult to think about friends who have learned to play music by ear, but who cannot give an account of the transitions in terms of written music language. That is to say that the phenomena can exist independently of being described in a particular code. It seems to us that this is the case with the experience of learning and using mathematics. The transition exists, but we still have to develop a common code and language to provide a more explicit account of the phenomenon.

Regarding the meanings of transition in psychological theory, as already mentioned above, the notion of transition in developmental psychology has been closely associated with stage theories. The concept was employed to contrast theories that saw mental growth as 'a cumulative affair, that new skills can be added steadily without modifying old skills in any significant way' (Miller, 1962, p. 319) with theories that saw child development as 'a series of abrupt transitions from one fairly stable stage to another which is equally stable but more advanced and (presumably) more complex' (Miller, 1962, p. 319).

Apart from this particular use in stage theories, rarely is the concept found in the subject index of general developmental psychology books. One of the key criticisms of stage theories was the lack of clear accounts of the passage from one stage to another, i.e. how the transitions took place. An exception to this type of treatment of the concept of transition appeared in Urie Bronfenbrenner's (1979) perspective on the ecology of human development. He proposed that a scientific theory on human development must consider the following three key features:

- the developing person, not 'merely as a tabula rasa on which the environment makes its impact, but as a growing, dynamic entity that progressively moves into and re-structures the milieu in which it resides';
- the interaction between person and environment as two-directional;
- that the environment relevant to developmental processes cannot be 'limited to a single, immediate setting but is extended to incorporate interconnections between settings'.

He suggested that the developing person is constantly experiencing movement in his/her ecological space, and based on this, he introduced the concept of transition. For him: 'an ecological transition occurs whenever a person's position in the ecological environment is altered as the result of a change in role, setting, or both' (p. 26).

Although he acknowledged that transitions are related to both biological and environmental changes, his theory departs from the mainstream developmental psychology by exploring the latter changes. Among various examples he gave of transitions the following are closely related to the focus of our book: entry into school, and emigrating to another country (or perhaps just visiting the home of a friend from a different socio-economic or cultural background) (p. 27).

However, Bronfenbrenner did not restrict transitions to movement between physical contexts. He also distinguished between ***role transitions*** in which 'the setting remained constant, but the subjects were successively placed in different roles, with corresponding alterations both in their own behaviour and in their treatment of others' (p. 103) and a ***setting transition*** which 'occurs when the person enters a new environment'. The notion of role transition is interesting in the sense that it allows for modifications to occur in the negotiations within a particular setting. Although it seems to us that the author used the notion of role in quite a static way, such as to mean the role of the pupil, the teacher, etc. it can be adjusted to incorporate the idea that roles are not just pre-givens, but are also mutually re-constructed in interactions.

The way Bronfenbrenner conceptualised setting transitions is also very illuminating. He spoke of dyadic relationships and their role concerning the nature of the activity or social setting experienced within and between ecological settings. In his view the setting transition needs to be understood not only in terms of what he named *primary links*, that is the experience of the person that is entering a new setting. In addition he argued for the need to consider the *supplementary links* which refer to other people related to the 'primary actor' and who can share the experience in forms that are more or less direct. For instance, if the primary link is related to a child going to school, parents can establish supplementary links either directly by going to school and meeting the teacher, or indirectly by relying on information brought by the child.

Twenty years have passed since Bronfenbrenner proposed his theory of development-in-context. Reading his book we gain the impression that he possessed rather static images of the transition processes compared to the ones we currently need. However, there is no doubt that his idea that development needs to map the experience of interconnections between contexts, some of which are experienced directly and others symbolically through others, is still very up-to-date.

A different approach to transition is offered by Beach (1999), also coming from a developmental psychological tradition but with strong sociocultural orientations. He is interested in developing more productive ideas about what used to be referred to as 'transfer' in learning. He has coined the term 'consequential transition' and has identified four main types: lateral, collateral, encompassing, and mediational, where 'Lateral and collateral transitions involve persons moving between pre-existing social activities. Encompassing and mediational transitions have persons moving within the boundaries of a single activity or into the creation of a new activity' (p. 114).

His typology will be very useful in this book, so let us clarify his categories of consequential transitions.

1. **Lateral transitions** – occur when an individual moves between two historically related activities in a single direction, such as moving from school to work. Participation in one activity is replaced by participation in another activity in a lateral transition.

2. **Collateral transitions** – involve individuals' relatively simultaneous participation in two or more historically related activities, such as daily movements from school to home.
3. **Encompassing transitions** – occur within the boundaries of a social activity that is itself changing, and is often where an individual is adapting to existing or changing circumstances in order to continue participation within the bounds of the activity.
4. **Mediational transitions** – occur within educational activities that project or simulate involvement in an activity yet to be fully experienced.

Beach (1999) argues that these four types of transitions share a common set of features, which are of strong interest to us in this book. Each potentially involves the construction of knowledge, identities and skills, or transformation, rather than the application of something that has been acquired elsewhere. Each is best viewed as a developmental process. Each involves changes in identity as well as knowledge and skill. Finally, each changes the relations between persons and social activities represented in various systems of artifacts, which as Beach says: 'This not only acknowledges the recursive relation between persons and activities, but makes it the explicit object of study' (p. 119).

Concerning sociological research in education transition is a concept very often used. In this context it tends to focus on how individuals move from one social institution to another. In the past some of these studies tended just to look at the products of transition. That is to say characteristics of one institution were treated as levels of an independent variable, e.g. children from single-parent families, and adjustment to the other institution as the dependent variable, e.g. level of success in school education. However, a look at the REGARD data base, which contains the summaries of projects funded by the Economic and Social Research Council (ESRC) in the UK, including also projects with an European dimension, shows a marked re-definition of the aspects of the transition needing to be analysed (ESRC, 1998–2000).

This shift is evident in the description of the foci of the projects. For instance, Brown, McNamara, and Hanley (1997–1998) proposed as their research objectives: providing a 'detailed accounting of the student's experience of moving from being a school student of mathematics to being someone who teaches the subject themselves' and 'a theoretical account of how notions such as progression, transition and development are constructed in accounts of the process of teacher training…'.

Evans, Behrens and Rudd (1998–2000) were involved in a project entitled 'Taking control' which 'aims to understand how young adults *experience control and exercise personal agency* as they pass through extended periods of transition in education and training, work …' They take as a key assumption that young people's 'experiences and their futures are not exclusively determined by socialising and structural influences, but also involve elements of *subjectivity, choice and agency*' (Our italics).

It seems to us that the current use of the transition concept in Beach's work and in other recent sociological and educational research provides exactly the type of dynamism that was lacking in Bronfenbrenner's framework. The key words to characterise the experience of transition as involving construction, control, subjectivity, choice, and agency are akin to current developments in cultural psychology (see chapter 7) in its attempts to provide a more adequate account of interface between sociocultural and psychological processes.

2.3. THE MEANING OF TRANSITIONS IN OUR BOOK

Based on the brief review presented above we can see firstly that the notion of transition is used in developmental psychology and education to mean movements that can occur between 'developmental periods'. For instance the movement from adolescence to adulthood. It is also used to mean transitions between contexts of practices, or major cultural institutions – such as from home to school (Corsaro, 1996).

Secondly, we can see that in the projects under progress the notion tends to be used in a dynamic way. That is, the way the person will experience a transition is viewed as involving input from both individual and sociocultural structures. Both types of inputs are then conceptualised as constantly evolving. Personal agency in the individual is paralleled by active social dynamics, such as inputs from significant others that are constantly negotiated in social interactions.

Bronfenbrenner's (1979) notion of role transition viewed from a more dynamic perspective will resonate with current views that learning within a practice involves a movement from the periphery to the centre (Lave & Wenger, 1991). Changes in roles could be closely associated with levels and types of engagement. It is expected that teachers as the official 'old-timers', and more experienced peers as the 'unofficial' ones, will play a major mediating role in the transition of the 'new-comers' to the classroom mathematical practices. However, following Bronfenbrenner these changes in roles within a practice cannot be conceptualised without also taking into account the setting transitions, i.e, the contexts of mathematical practices. It may be that salience of movement within and between practices varies according to points in life trajectories of learners. This, however, is an empirical question, which we hope the empirical research reported in the book will help us to start addressing.

In our understanding, the contributors to this book chose to approach transition according to a dynamic perspective. Although we hope the research will give us some insights into role transitions necessary for progressive movement within a practice, our main interests lie in clarifying the experience of transitions between contexts of *learning* and *using* mathematics; between contexts for different mathematical practices. In our work the reference to contexts sometimes acquires a status of site, which can be physically located. Home, school, work, after-school tutorials, 'Saturday' schools, are sites where (socio-culturally) organised mathematical practices take place. However, for a learner in transition between contexts, one of the contexts can become an invisible but present structure. What will count in this case will be the knowledge, tools, values and meanings that the learners in transit evoke or suppress to cope in the context where they are physically present. What we are interested in is understanding the dynamics of this process.

Expanding on Bronfenbrenner's notions that transitions can occur between settings or between roles, and using Beach's categories of transition, we will consider in this book various aspects of the transition experience arising from:

- Lateral transitions, where individuals move between two related practices in a single direction. The situation of immigrant students in mainstream schools is an example of this type of transition, where the students are reconstructing their identities in their new country. Subsequent chapters will consider varying perspectives, such as the students themselves (Gorgorió et al., Bishop), their parents (Abreu et al.; Civil & Andrade) or their teachers (Civil & Andrade). Santos and Matos' chapter also relates to the school-to-work example of lateral transition.
- Collateral transitions, where there are two or more related practices requiring relatively simultaneous involvement. Again subsequent chapters will consider for example the situation where the school students' parents emigrated after being at school in their home country, and the student is exposed to one set of mathematical practices and representations at home and another set at school. Examples of these transitions will be given in the chapters by Abreu et al., Bishop, Civil and Andrade and Gorgorió et al..
- Encompassing transitions, where the individuals or groups experience a significant change in mathematical practices due to historical changes within their own developing institutions or communities of practice, giving rise to cognitive and social conflicts. This type of transition occurs in circumstances such as with the white-English parents supporting their children's school mathematical learning (Abreu et al., chapter 5), with the teachers in the Civil and Andrade study, and with the *ardinas* in Cabo Verde, who moved from school to selling newspapers, but who now experience changes in the selling practice when the newspaper organisation changes (Santos & Matos, chapter 4).
- Mediational transitions, where the individual or group interacts in an intentionally educational activity designed to change perceptions and meanings before involvement in an activity yet to be fully experienced. As illustrated in Civil and Andrade's chapter the interactions that take place between women's groups in the mathematical workshops, for example, and the questions that the women raise are facilitated by the dyads and triads of women that come together to learn. They enter a number of social settings together, where they might not otherwise venture.

3. OVERVIEW OF THE BOOK CHAPTERS

Following this chapter five empirically focussed chapters explore transitions at in-school and out-of-school research sites, and although each of the authors and research groups have used related theories to inform their work, the approaches and constructs they have used are quite varied. We revisit these constructs in the more reflective chapters seven to nine in order to explore how they could contribute to a further development of the research.

In Chapter 2, Núria Gorgorió, Núria Planas and Xavier Vilella describe the processes and challenges of establishing and carrying out a research study on immigrant children in schools in Catalonia, in North East Spain. Their chapter reveals the complexities of doing this kind of research, complexities which are not just in terms of methodology but which reflect political and social aspects of transition research. They also report on and analyse the sociocultural conflicts that the students experience in their mathematics lessons as well as the teachers' understanding of those conflicts. Their analysis is of transitions being negotiated at a microgenetic or face-to-face interaction level.

Chapter 3 by Alan Bishop also focuses on the situation of immigrant students, and he reports on a study undertaken in Australia with students classified by the researchers as having a Non-Anglo-Cultural Background (NACB). Such students are often recent immigrants to Australia trying to adapt to a new educational situation, and the author analyses their transition experiences through the quantitative and interview data obtained in the study. The study demonstrates the range of influences on the immigrant students, together with the dilemmas and decisions they face in coming to terms with their new learning context.

In Chapter 4, Madalena Santos and João Filipe Matos describe their research on the transitions experienced within an out-of-school mathematical practice in Cabo Verde involving young newspaper sellers. They use this research to elaborate Lave and Wenger's notion of 'learning as increasing participation in communities of practice', and explore it from two perspectives (a) the socio-historical organisation of the social practice and, (b) the sellers (*ardinas*) as individual participants in the practice. Evidence of living transitions emerges in the descriptions of how *ardinas* act to sustain their participation in the social practice. They move from being newcomers to old-timers, from one role to the other, between different rules, values and discourses. Santos and Matos argue that this understanding of living transitions that sustain individuals and communities are also relevant to the formal mathematical practices of school.

In Chapter 5, Guida de Abreu, Tony Cline and Tatheer Shamsi report on a study involving the parents of children attending multiethnic primary schools in the South of England. The chapter addresses the relevance of taking into account the roles played by parents in the way children experience their school mathematical learning. By visiting and interviewing parents the authors obtained the accounts of how they participated in their child's mathematics learning. These accounts were very illuminating in the way that they highlighted how cultural backgrounds contribute to make salient particular aspects of the experience. However, in terms of understanding the dynamics of the process of transition, as experienced by particular children, the authors obtained deeper insights when they took into account the interaction between the parent and the child's perspectives.

In Chapter 6, Marta Civil and Rosi Andrade offer a number of transitions in progress. One transition analyses changes in teachers' beliefs and practices as they strive to bring the community into their classrooms. In study groups, the teachers report their ethnographic household visits whereby they interview parents and others in the community and discuss the use of these practices for the learning of mathematics in the classroom. In some cases, the teachers invite the parents to visit the classroom and work with the students to share the mathematical ideas involved in the practice. Another transition offers insight into mothers' experiences as learners of mathematics themselves, with the intent of becoming more informed in their educational support at home. And yet another transition looks at children's perceptions of out-of-school and in-school mathematics.

There next follows a brief introduction to the following three chapters that each reflect on different theoretical perspectives concerning transition research. They discuss at a meta-level the issues of researching transitions, particularly how and where one situates the research and they draw on the results and analyses reported in the previous five chapters. There have been several important developments in theory as a result of the increasing convergence of research interests in mathematics education, sociocultural psychology, anthropology, sociology, linguistics, semiotics, political and gender studies. These three chapters focus on developments that seem to hold particular promise in grappling with the complexities of mathematical transitions.

Chapter 7 is by Guida de Abreu and explores the contribution of a cultural psychology perspective to the understanding of how a person experiences transitions between different contexts which require either learning or uses of mathematical knowledge. In this chapter Abreu suggests that the study of transitions between practices represents a third wave in the way relationships between learning and cultural practices are examined. The chapter starts with an outline of the notions of 'cultural practices' and 'cultural-tool mediation' in the cultural psychology approaches associated with Vygotsky's theory. Next, drawing on empirical research she illustrates how a focus on transitions foregrounds the importance of the valorisation of specific practices by societies, institutions and communities of practice. She argues that the value-oriented aspects of mediation need to be examined as a complementary aspect of cultural-tool mediation. Finally, the concepts of 'identity chaining' (Valsiner, 2000), 'competing identities' and 'projected identities' are introduced to explore psychological issues related to the impact of valorisation of mathematical practices on the person involved in transitions.

In Chapter 8, Alan Bishop focuses on the anthropology-based constructs of mathematical acculturation and cultural conflict and develops these ideas in the context of the empirical studies reported in the earlier chapters. The chapter addresses the co-construction of the transition process by the significant players in the institutional and social contexts, and thereby expands the constructs of conflict and power. Bishop also explores the relevance of the ideas of the mathematics classroom as a workplace, and the borderland discourse found there, together with the potential of the idea of cultural production as being a more productive mediational construct than acculturation in situations of transition.

Chapter 9, by Norma Presmeg, addresses themes in the shifting meanings that participants may attribute to their experiences during transitions between contexts. In all of the empirical chapters of this book, mathematical meanings with their symbolisations are intertwined with sociocultural meanings in the experiences of those participating in transitions. In many cases these personal meanings influence the construction of the self-images – the very identities – of those involved. Linguistic competence, psychological coping mechanisms, power relations, shifting value systems influencing what counts as knowledge, are some of the elements that may be involved in the changes in mathematical meanings that are part of transitioning. This chapter details some of the dynamics of the back-and-forth movement in which current mathematical experiences are interpreted on the basis of meanings and representations acquired in previous situations.

Finally in the Postlude, the main overarching issues to have emerged from the diverse chapters in the book are summarised and some pointers to future needs in research are discussed. In particular the notion of mediation in transition situations, which emerges in each of the three reflective chapters, is discussed.

REFERENCES

Abreu, G. de (1995). Understanding how children experience the relationship between home and school mathematics. *Mind, Culture and Activity, 2*(2), 119–142.

Abreu, G. de (1998). *Reflecting on mathematics in and out of school from a cultural psychology perspective.* In A. Olivier & K. Newstead (Eds.), *Proceedings of the 22nd Conference of the International Group for the Psychology of Mathematics Education, Vol 1* (pp. 115–130). Stellenbosch, RSA: University of Stellenbosch.

Abreu, G. de, Bishop, A., & Pompeu, G. (1997). What children and teachers count as mathematics. In T. Nunes & P. Bryant (Eds.), *Learning and teaching mathematics: an international perspective* (pp. 233–264). Hove, East Sussex: Psychology Press.

Apple, M.W. (1998). Markets and standards: the politics of education in a conservative age. *Proceedings of the 22nd Conference of the International Group for the Psychology of Mathematics Education, University of Stellenbosh, South Africa, 1*, 19–32.

Barton, B. (1996). Anthropological perspectives on mathematics and mathematics education. In A.J. Bishop, K. Clements, C. Keitel, J. Kilpatrick & C. Laborde (Eds.), *International handbook of mathematics education* (pp. 1035–1053). Dordrecht: Kluwer.

Beach, K. (1999). Consequential transitions: a sociocultural expedition beyond transfer in education. *Review of Research in Education*, 24, 101–139.

Bishop, A.J. (1988a). *Mathematical enculturation: a cultural perspective on mathematics education.* Dordrecht: Kluwer.

Bishop, A.J. (Ed.). (1988b). *Mathematics, education and culture*. Dordrecht: Kluwer.

Bishop, A.J. (1993). Influences from society. In A.J. Bishop, K. Hart, S. Lerman & T. Nunes (Eds.), *Significant influences on children's learning of mathematics* (pp. 3–26). Paris: Unesco.

Bishop, A.J. (1994). Cultural conflicts in mathematics education: developing a research agenda. *For the Learning of Mathematics*. 14, 2, 15–18.

Bishop, A.J. & Goffree, F. (1986). Classroom organisation and dynamics. In B. Christiansen, A.G. Howson and M. Otte (Eds.). *Perspectives on mathematics education*. Dordrech: Reidel.

Bourdieu, P. (1995). *Language and symbolic power.* Cambridge, Massachusetts: Harvard University Press.

Bronfenbrenner, U. (1979). *The ecology of human development*. Cambridge, Mass.: Harvard University Press.

Brown, A.M., McNamara, O., & Hanley, U. (1997–1998). *Primary students teachers understanding of mathematics and its teaching*. Manchester: Department of Science Education, Manchester Metropolitan University http://www.regard.ac.uk/.

Bruner, J. (1990). *Acts of meaning*. Cambridge, Mass.: Harvard University Press.

Bruner, J. (1996). *The culture of education*. Cambridge, Mass.: Harvard University Press.

Cobb, P. (1995). Mathematical learning and small group interaction: four case studies. In P. Cobb & H. Bauersfeld (Eds.), *The emergence of mathematical meaning* (pp. 25–129). Hillsdale, New Jersey: Lawrence Erlbaum.

Cobb, P., & Bauersfeld, H. (1995). *The emergence of mathematical meaning*. Hillsdale, New Jersey: Lawrence Erlbaum.

Coombs, P.H. (1985). *The world crisis in education: the view from the eighties*. New York: Oxford University Press.

Corsaro, W. (1996). Transitions in early childhood: the promise of comparative, longitudinal ethnography. In R. Jessor, A. Colby, & R.A. Shweder (Eds.), *Ethnography and human development* (pp. 419–457). Chicago: The University of Chicago Press.

Duveen, G. (1998). The psychosocial production: social representations and psychologic. *Culture & Psychology, 4*(4), 455–472.

ESRC. (1998–2000). *Youth, citizenship and social change* : http://www.tsa.uk.com/YCSC/Intro.html.

Evans, K.M., Behrens, M., & Rudd, P.W. (1998–2000). *'Taking control': agency in young adult transitions in England and new Germany*. Surrey: University of Surrey, School of Educational Studies, http://www.regard.ac.uk/.

Goodnow, J.J. (1990). The socialization of cognition: what's involved? In J.W. Stiegler, R.A. Shweder, & G. Herdt (Eds.), *Cultural psychology* (pp. 259–286). Cambridge: Cambridge University Press.

Lave, J., & Wenger, E. (1991). *Situated learning and legitimate peripheral participation*. Cambridge: Cambridge University Press.

Lerman, S. (Ed.). (1994). *Cultural perspectives on the mathematics classroom*. Dordrecht: Kluwer.

Miller, G.A. (1962). *Psychology: the science of mental life*. Harmondsworth: Penguin Books.

Moscovici, S., & Paicheler, G. (1978). Social comparison and social recognition: two complementary processes of identification. In H. Tajfel (Ed.), *Differentiation between social groups: studies in social psychology of intergroup relations*. London: Academic Press.

Nunes, T., Schliemann, A., & Carraher, D. (1993). *Street mathematics and school mathematics*. Cambridge: Cambridge University Press.

Pinxten, R. (1994). Anthropology in the mathematics classroom. In S. Lerman (Ed.) *Cultural perspectives on the mathematics classroom* (pp. 85–97). Kluwer: Dordrecht

Planas, N., Vilella, X., Gorgorió, N. & Fontdevila, M. (1999). Fiayaz en la clase de matematicas: ambiente de resolucion de problemas an un aula multicultural. *Revista sobre la ensenanza y aprendizage de las matematicas, 30*, 65–75.

Presmeg, N.C. (1988). School mathematics in culture-conflict situations. *Educational Studies in Mathematics, 19*(2), 163–177.

Presmeg, N.C. (1998). A semiotic analysis of students' own cultural mathematics. In A. Olivier & K. Newstead (Eds.), *Proceedings of the 22nd conference of the International Group for the Psychology of Mathematics Education, Vol 1* (pp. 136–151). Stellenbosch, South Africa: University of Stellenbosch.

Secada, W. (1992). Race, ethnicity, social class, language, and achievement in mathematics. In D.A. Grows (Ed.), *Handbook of research on mathematics teaching and learning* (pp. 623–660). New York: Macmillan.

Siegler, R.S. (1996). *Emerging minds: the process of change in children's thinking*. Oxford: Oxford University Press.

Valsiner, J. (2000). *Culture and human development*. London: Sage.

Van Oers, B., & Forman, E. (1998). Introduction to the special issue of learning and instruction. *Learning and Instruction, 8*(6), 469–472.

Walkerdine, V. (1988). *The mastery of reason: cognitive development and the production of rationality*. London: Routledge.

Wertsch, J.V. (1991). *Voices of the mind*. Cambridge: Cambridge University Press.

CHAPTER 2

IMMIGRANT CHILDREN LEARNING MATHEMATICS IN MAINSTREAM SCHOOLS

NÚRIA GORGORIÓ, NÚRIA PLANAS AND XAVIER VILELLA
Universitat Autònoma de Barcelona

1. IMMIGRANT CHILDREN LEARNING MATHEMATICS IN MAINSTREAM SCHOOLS: A TRANSITION PROCESS

We understand the schooling of the immigrant[1] children as a transition process, because when they arrive into a new country they have to cope with the many changes involved in moving from one culture to another. In particular, they have moved from one school culture into another, if they have attended school, or perhaps they have moved from a 'no-schooling' culture into a school culture. We regard immigrant students as having the need to build a bridge from the meanings of their initial situation to those of the present one. All of them have the right to be offered the opportunity to develop their potentialities to the full, regardless of their country of origin or the reasons for their migration. We believe that school should contribute to help them create a continuity between their home and the host culture's meanings. From that point of view, and not avoiding the researchers' commitments to society and particularly to teachers and students, the goal of our study is to find teaching approaches that contribute to co-construct the students' transition in order to make it as smooth as possible.

We say 'co-constructing the transitions' because a one-sided construction would not be complete, since the meanings a child brings to a situation, as Bruner states, 'are not to his own advantage unless he can get them shared with others' (1990, p. 13). Everyone involved in the dynamics of the mathematics classroom has to participate in the negotiation of the meanings associated with the diverse situations, in order to ensure a real sharing of them (Kao & Tienda, 1998).

[1] In our study, the word 'immigrant' is considered as taken in Ogbu's classification of minorities (Ogbu & Simons, 1998). Therefore, we take into account voluntary immigrant minorities, who are supposed to have moved willingly to Catalonia, and involuntary immigrant minorities, such as refugees, migrant/guest workers, undocumented workers, and binationals, including descendants or later generations. Even though there can be different types of minority status among these groups, all of them have in common, to some extent, the need for a social adjustment and equal educational opportunities in their school performance.

G. de Abreu, A.J. Bishop, and N.C. Presmeg (eds.), Transitions Between Contexts of Mathematical Practices, 23–52.

Smoothing the student's transition would require, in particular, making the cultural conflicts something positive, both for immigrant and local students. When analysing the meanings that a child brings to a school situation, it has to be taken into account that those are constructed in relationship not only with the sociocultural context of the learning, but also with his or her emotions, values and beliefs. Immigrant students cannot establish reference points with which to direct themselves, without the guidance and the acceptance of people belonging to the host culture, the teacher and their classmates, among their 'significant others'.

In this chapter we present the analysis of the immigrant students' process of transitions, from their home and school culture to the school culture that hosts them, and we shall focus on the mathematics classroom, and understand the construct 'culture' in its broadest sense. By focusing on the cultural conflicts that arise in a mathematics classroom, we study the transition processes both as they are understood by the teachers and as an external manifestation of how the students themselves adapt to the changes, by constructing new meanings and values and adapting the old ones. In particular, we will refer to the social and sociomathematical norms, and the norms of the classroom mathematical practices.

We take as our starting point the definition of culture by Geertz (1973) and we analyse the context of the transition process, being predominantly the mathematics classroom, but we consider it within the school, and within the educational and social structures that frame it, and condition what is possible and what is desirable. We consider the main interest of our project to be knowing more about how the significant persons that influence the learning process as a transition process, essentially the teachers, understand these processes. As our research is not only interpretative, but also has the intention of promoting change, we have been designing, experimenting with, and analysing, different classroom situations that are potentially useful for making explicit and positive the cultural conflict which is often invisible in the mathematics classroom.

We understand the construct 'transition' not as a moment of change but as the experience of changing, of living the discontinuities between the different contexts, and in particular between different school cultures and different mathematics classroom cultures. The construct 'transition' is, in our understanding, a plural one. Transitions arise from the individual's need to live, cope and participate in different contexts, to face different challenges, to take profit from the advantages of the new situation arising from the change. Transitions include the process of adapting to new social and cultural experiences, and students need to be helped to understand the meanings of the new experiences and to reinterpret them and construct new ones based on their own individual meanings and values.

Researchers and teachers can only see the external part of the transition process and they only have the means to interpret it. As it is a private and personal process, and most of the time hard to exteriorise, one can just interpret what is going on in the student's transition process through its external manifestation. The mathematics classroom is a social and cultural scenario and, as in every educational situation, it has its social dynamics and its social interactions. The various moments of those dynamics have different meanings for the different participants in them and these differences can create cultural conflicts (Bishop, 1994). On the other hand, as Bishop remarks, there is an unavoidable part of cultural conflict in every educational situation.

Cultural conflicts and disruptions between the various meanings that different persons attach to the same situation are, probably, the most visible manifestations of the transition processes lived by the participants in a multicultural classroom. However, misbehaviour, lack of interest, absenteeism, could be reinterpreted as an external manifestation of a difficult transition process, if the observers were aware of the tensions lived by the students in the new situation. Often, the apparent lack of conflict only means its invisibility to the observers, and when cultural conflict remains invisible it may turn into different types of blockage that can slow or hinder immigrant students' learning process and their participation in classroom community life.

Researchers and teachers also have their own meanings and expectations related to classroom situations. Thus, for instance, teachers find immigrant students to be 'different' from what they expect their pupils to be. When talking about differences in a social situation we mean differences from the 'normality', where this is defined according to the assumptions and expectations of the persons concerned. As Bauersfeld et al. (1988) state: 'social interaction takes place among individuals or subjects, which mutually constitute expectations, interpretations from each other, and test these interpretations by negotiation processes, producing, this way, meanings, structures and acceptation norms and norms to validate' (Bauersfeld et al., op. cit. p. 174). When acknowledged, cultural conflict can be assumed as a positive starting point in order to accept the fact of cultural diversity, and making it explicit is the first step to facilitating the students' transition processes. Therefore, for a real sharing of meanings, it is important that the adults involved in the teaching of immigrant students are explicitly conscious of their own. More than that, they should be ready to review them and to change them if they want their students' transitions to be co-constructed: the move has to take place on both sides.

The immigrant students that we have worked with experience different kinds of transition processes. Some transitions are go-and-come-back continuously, following the classification given in chapter 1 we can call them 'collateral transitions', where students participate in the experience of more than one context, for instance, the mathematical practice inside and outside school. These transition processes should contribute to give plural meanings to the signifiers and to mathematical knowledge. The students also experience 'lateral transitions', the transitions resulting from an irreversible change, if not psychological, at least physical, having moved from one country to another where they now have to live. We understand this transition process as being most significant when establishing their path of progress, since it is linked to opportunities and barriers.

However, what is important about transitions, all of them, is that the immigrant students move from a world with particular meanings and values to another world with other meanings and other values. To be able to react to them, by appropriating them, or not, they need to understand and reinterpret them both on the basis of the meanings and values they had in the previous context and on those they perceive in the new situation. It is also important to take into account that the more difficulties the individuals have to structure the new meanings, the more obligation there is for us as educators to help them to create a continuity bridge. The 'good' students, from a high social and cultural class, even if they are immigrant students, have fewer difficulties in engaging in the process of transition and thus suffer less. The unschooled students, with social, familial and economic difficulties, need to be helped much more to 'organise and structure' their new meanings. The more distant the meanings are from the different worlds, the more need there is for making explicit those in the new situation.

The issue then, in the context of our project, is what are the continuities and discontinuities, their coherence and non-coherence, between the meanings attached to the mathematics learning process by the different participants in it? What are the meanings that teachers and students attribute to learning mathematics, to the different mathematical practices within and outside school, to successful learning, to assessment, to mathematical usefulness? And what are the values associated with these by the different participants?

The more we help to further the coherence, the smoother will the transition be, and the greater will be the opportunities for students to learn mathematics. We are not facing students defined only as 'being from another culture' or another country, but as students who are at a certain moment in time in a continuity between the two cultures. It is one of our goals to try to find ways to help the teachers to contribute to creating this continuity and coherence in the entire educational activity. Continuity cannot be established without the clear intention of acknowledging the student's culture and the culture of the group. History is full of abrupt breakdowns of this continuum, and of curricula imposed artificially that are far from the real needs of the individual.

In our understanding, coherence and continuity are not only necessary from the point of view of educating or helping in the development of every immigrant student as a person, but are 'powerful' social tools. Transitions between different educational cultures have certainly to do with issues of equity and justice. Transitions are related to social progress, and have to do with 'social selection'. Transition is not a matter of 'changing the scenery or the decor' of the educational process but it is about living changes that are linked with chances of success. There are transition processes that are more likely to result in the child succeeding at school and, from the point of view of the system, a 'successful transition process' would be the one that enables the student to get 'good results' within the system. However, the transition process must also be a positive one for the person, one that is lived as enrichment. What for some people will be a benefit, for others could be a loss. Furthermore, we would argue that it should be a process through which people adapt to the new situation without having to give up their cultural background, but can reinterpret it in the light of their present needs.

Mathematics educators, teachers and researchers, through their attitudes perpetuate the myth that the subject is just for elites, consciously confirming the failure of some students through poor learning conditions, or unwittingly through prejudice, values and expectations (Apple, 1998; Dowling, 1998). Researchers and teachers should be aware that teaching may induce, consciously or not, intentionally or not, failure among the immigrant students through their transitions. In their individual action as agents for change they have a big responsibility that cannot be avoided or excused by the constraints of the structure that limits it.

2. THE RESEARCH CONTEXT: ITS COMPLEXITIES

In recent years, there has been an increasing immigration into Catalonia, an autonomous region in northern Spain (whose capital is Barcelona), which has led to significant changes in the school population. The immigrant population in Catalonia is about 1.4% of the whole population and, in 1997, it reached 2.3% in Barcelona. This percentage is not homogeneously distributed in the city; in particular, in the area where we have focussed our research the percentage of immigrant students rises to 90%. This new situation has focused attention on the inadequacy of the educational provision in multicultural schools and classes and raises many questions related to issues of equity and justice.

In 1997, the first of the authors received a grant from a Catalan private foundation devoted to education, Fundació Propedagògic, to carry out a project concerned with mathematics teaching in schools that have large numbers of immigrant students[2]. The project was also supported by the Catalan Ministry of Education. In this section we introduce the research context and we discuss its complexities, both from the global point of view, as to how the political and social structures are in tension with the researchers' assumptions, and from a more local point of view, the complexities of the classroom reality in everyday life.

Even if the project was initially linked to a request from the administration, the team's understanding of the multicultural situation in schools goes far beyond that of the educational administration. The team negotiated strongly to change what initially was expected to be a policy-driven 'research' project into a research project with no inverted commas. Probably, the most difficult argumentation with bureaucrats and politicians, has been about mathematics being a cultural product and that learning and teaching mathematics is linked to values, beliefs and expectations and that this emotional aspects can explain many of the difficulties immigrant students have when learning mathematics.

To acknowledge mathematics as a cultural product is a first step to taking advantage of the cultural diversity among the students as a source of richness for mathematics learning (Wilson & Mosquera, 1991). On the other hand, since any mathematics classroom can be considered to be a multicultural class, understanding culture in a broad sense (Borba 1990), an approach that considers mathematics as a cultural product will benefit all students, whether they are immigrant or not. The search for curricular models and methodological approaches that take culture into

---

2 The context of the research has already been presented as a contribution to MEAS1, 1st International Conference on Mathematics Education and Society, held at Nottingham on September 98: Gorgorió (1998)

account has as a main goal to facilitate both the 'enculturation' and the 'acculturation' processes (see Bishop's chapter 8, this book) within the mathematics classroom.

We have also spent part of our time in trying to change the educational bureaucrats' idea of a 'cognitive or cultural deficit' to be 'compensated', by the importance of building on the potentialities every student has. The authors consider that explaining the difficulties immigrant children have in our schools in terms of cognitive deficit is too simplistic and questionable (Ginsburg & Allardice, 1984; Nunes, Schliemann & Carraher, 1993; Rasekoala, 1997). Moreover, this interpretation has social implications because it projects particular expectations onto concrete cultural groups, a fact that confirms our feelings against it because of our personal values and experiences. We understand, as Crawford (1986), that the difficulties immigrant students experience when learning mathematics are often linked to the distance of their different social and cultural frames of reference from the implicit ones within the school. Our starting point is to consider the cultural contribution of ethnic minorities and of the different social groups as a source of richness to be maintained and shared. The team regards cultural differences and the cultural conflicts arising from them as a potentiality, not as a 'problem to be solved' nor as a 'diversity to be treated', as it was considered by the educational administration, school inspectors and principals.

We have to deal with students whose parents belong to a culture different from the one that hosts them, but we regard the students themselves as being at a certain point in a continuum between their familial culture and the host culture. Therefore we believe that the main educational approach should be to help them create their own psychological and social identities (Abreu, 1995). In line with this belief, we want our research and its implications to take into account the students' out-of-school knowledge, including their values, beliefs and expectations. But the question for us was how to take this knowledge and these values into account, especially because the initial request from the administration was to create 'ready-to-use' materials, edited in different languages, that could be given to teachers having immigrant students in their classes. We see no point, for example, in trying to teach aspects of the Moslem history of mathematics when either the students make explicit in class that they want to become fully integrated Catalan adults or hide their origin by disguising their first names as Catalan ones.

The broader social structures where any educational act is embedded, limit and restrict what is possible to change and how it can be changed. Every research project is a situated research project and, if one wants it to be potentially useful in other contexts, there is the need to make explicit the context where it is framed, with its constraints, possibilities and challenges. The uniqueness of the context conditions greatly not only the results, which are certainly not general, not only the methodological procedures that can be used, but also and mainly the goals that can be established in a sensible way. The results of any research in mathematics education relate to a previous 'problematique' and to the methodology chosen in a particular situation. Therefore, in the next paragraphs we will briefly describe the highly politicised sociocultural context, within the legal and institutional frameworks.

The legal regulations concerning foreigners in Spain reflect the social reality concerning immigration. The implications of the immigration legislation for the educational situation of immigrant children have already been discussed elsewhere (Gorgorió, 1998). However, we want to point out here that even if the law considers the regrouping of the family as a possibility, the practical and legal obstacles needing to be overcome to achieve it lead to the presence of many unstable, and unstructured families. The anxieties and tensions that immigrant families face are reflected in the relationships that their children establish with the school (Nieto 1999).

The increasing number of immigrants since the beginning of the nineties, together with the economic and structural crisis, and the concentration of the immigrant working population in certain areas of the country, no longer creates the illusion that our society is a tolerant society, when the individuals concerned feel that their integrity or their status is at risk. From the point of view of the schooling of immigrant children, there is a crucial issue resulting partially from the above situation: many parents of Catalan children have moved them from the public schools which have 'too many' immigrant children into private schools. In schools with an enormous percentage of immigrant children, the acculturation into the Catalan community is difficult because the only reference models the students have are their teachers. Moreover, since immigration is often linked to economic deprivation and social risk in certain areas, what could be 'normal' schools turn out to be what we could name 'ghetto' schools.

By the end of the 1980s, both state and regional governments with educational power in Spain (as in the case of Catalonia) promoted a broad reform of the educational system. The implementation of the new Educational Act, LOGSE[3], is currently a fact. We can claim thereby that for the past decade, education to age 16 has been regarded as a right accessible to most children in the country. However, even if the different educational reforms tend to support the democratisation of learning, it is a fact that we are still far away from the utopian idea of 'mathematics for all'.

Generally speaking, the present system meets current educational needs much better than the previous one, although there is much to be done in the area of teaching in multicultural situations, which is one of the weakest points in the implementation. The official regulations are very general, because there is an explicit intention that every school should adapt the curriculum to the actual needs of their pupils, and to its own possibilities for meeting those needs. The official documents are of little help to teachers since they address the multicultural facts in a quite naïve way. Moreover, old beliefs have not yet been abandoned by most of the educational community, particularly by principals and inspectors who do not consider cultural diversity to be an urgent issue. In addition, in-service teacher education programs dealing with multicultural education are scarce, and consequently the teachers and schools do not yet receive enough support to carry out any teaching innovations.

Considering the context of our project more locally, the diversity of students is one of the essential characteristics. Students from Magreb, India, Pakistan, China,

---

[3] LOGSE, Ley de Ordenación General del Sistema Educativo.

Japan, Gambia, Senegal, South America, among other origins, and the Catalan gypsy students, all of them with their rich cultural backgrounds, create in some mathematics classrooms a highly multicultural and multiethnic 'ambiance'. Apart from the students' place of origin, there are several other variables that contribute to the student diversity and that affect directly the social dynamics of the classroom. Therefore not only the process of teaching and learning mathematics, but also the possible research methods are also affected. Variables like gender, which in other contexts might not be considered so relevant, take on an enormous significance in our context. Gender differences are more acute in some cultures than in others, and this may create particular cultural conflict situations from the point of view of both teaching and researching. Male students who do not accept female students in their working groups, or girls who do not agree to be videotaped for the research, would just be some examples of such situations.

The various languages involved in the teaching and learning is another characteristic to be considered in the analysis of the students' transition processes, that moreover, at the same time, conditions the research approach and possibilities. In the multilingual settings that are the multicultural classrooms in which we have studied, teachers and students face linguistic distances that increase the communication gap in a double sense: the objective distance that exists because of the use of no common language, and the social distance in the meanings endowed to the messages once a common language has been acquired (see Gorgorió & Planas, 2000b, or Gorgorió & Planas, in press, for a full account of the analysis of teaching immigrant students from the point of view of language and communication).

The complexity of the research context creates a particular challenge arising from the need to balance the assumptions and the aims of the research as a socially committed project, the interests of the educational administration that supports it, and the respect towards the communities that participate in the study.

3. THE NEED FOR A MOVE: COLLABORATIVE RESEARCH

Having received the request to address an issue which is mainly connected with schools, our next point was to argue for a collaborative research team, to include both university researchers and in-service teachers. The dichotomy between research and practice is no longer acceptable. There is a need for a move to involve teachers in research, their roles consisting not only of developing the researcher's proposals, but being full participants in the whole research process. The need exists for knowing more about *teachers'* perspectives in practical issues which researchers could seriously address, and for counting on their expertise and knowledge to find ways to research them and to interpret the results (Bishop, 1996). This approach facilitates, more than any others, changes in the practice and the interiorizing of the research results and implications by those that are assumed to implement them.

The result of the negotiation between the first author, being the researcher commissioned for the project, and the educational administration, resulted in a collaborative team. The other members are in-service secondary and primary mathematics teachers already linked with research at university who have been granted partial release from their teaching hours to devote time to the project. With such a team we consider that not only do we have the support[4] and the expertise from the university level, but also knowledge and expertise from school practitioners, and even more importantly we are considering issues directly related to practice.

The reasons for deciding to develop the research collaboratively are several. We address a crucial issue related strongly with social demands within a particular context, and teachers are the ones who best know not only the significant and urgent needs, but also the possibilities and limitations of the system and the complexities of the social context. They know better the other practitioners who shape and constrain mathematics education, the school curricula, structure and timetable, and the familial contexts of the students. And it is also the teachers who know more about the real possibilities for change and its implementation.

Moreover, teachers' questions and explanations are from a knowledge domain that is distinct from and complementary to that of isolated researchers at university. Working collaboratively facilitates communication between the different domains, overcoming the mutual exclusion of practice and research, and also helps in finding ways to disseminate the research findings and the innovation proposals. The collaborative work allows us not only to take into consideration the factors that condition practice, but also the connections with published theory. Both of these play an important role in shaping the research, by establishing the possibilities, limitations and constraints of the context, and also by offering the dimensions of generality that give sense to research. This makes the whole study both an analysis of practice and a search for explanations towards a development of theory.

At a time when mathematics teaching is facing many tensions with the implementation of the new educational system in Spain, we consider that collaborative research can help the educational community to commit itself to the changes in educational practice, in terms of both the agents (researchers and teachers) and the structures (the educational administration). However, doing research within a collaborative model with in-service teachers, and having teachers participate in action-research projects implies a move that, at least in our context, still needs to be 'explained' and justified both to the university and to the school systems (see Gorgorió & Planas, 2000a).

Collaborative research also has its limitations, both practical and methodological. On the one hand, the teachers involved continue to do their jobs within the schools, and this means within the constraints of the administration, particularly regarding the time they are allowed to devote to the project. On the other hand, even if following that approach makes it easier for teachers to participate in research studies than with other approaches, some difficulties and tensions arise from the situation of teachers researching their own teaching (see Gorgorió & Planas, 2000a). However,

4 The members of the team want to take this opportunity to thank Guida de Abreu, Alan Bishop, Ken Clements and Norma Presmeg, for their support and their advice on starting the project and during its development. We want also to thank the Centre de Recerca Matematica, Institut d'Estudis Catalans, for having funded the TIEM98 project that gave us the opportunity to work with them during their stay in Barcelona.

from our understanding and in the context of our project, the outcomes we could achieve through a collaborative research approach are worth all the constraints and conflicts we are facing.

Regarding our action-research, we had to deal with the tensions between the teachers' responsibility to the students and to the research, particularly in relation to those issues of students feeling reluctant to participate, for instance, by not wanting to attend a class if it was going to be videotaped. Regarding this fact, we explicitly agreed that we had a communal responsibility as teachers above that as researchers, even if that could mean a 'loss' for the study, because we accepted this kind of limitation as part of our working with people that have their own system of values. Moreover, we did not consider this kind of situation as a 'loss' because it also gave us knowledge about the conflicts arising from the students living at one time in two cultures.

Phenomenological models (Eisenhart, 1988) insist on considering the educational phenomena throughout the real experiences of all those who are participants in it, therefore sociocultural research cannot be restricted to observation and measurements. Lerman (1996) also points out the limitations of a quantitative methodology in any analysis that tries to incorporate sociocultural variables and the affective dimension of learning. He points out the difficulties of dealing with multi-dimensional phenomena, like those taking place within the classroom, with uni-dimensional tools, which are usually used in quantitative methodologies. In general, all the studies that consider the social and cultural context of the learning have required a qualitative methodology (Bishop, 1988). The studies that consider the cultural aspects of all educational phenomena have consolidated this paradigm in the field of mathematics education (Schoenfeld, 1994). As an example, we could consider Cobb's perspective (Cobb, 1989) that justifies the use of a qualitative methodology in research where the social context and interactions are prior aspects when constructing meanings.

Qualitative methodologies are used in the studies that search for links between the person, the immediate context and the more global context where it is embedded. Eisenhart (1988) suggests a qualitative methodology when approaching a comprehensive theory that links cognitive and sociocultural theories. Abreu (1993) states that an ethnographic qualitative approach is useful in studies within classroom communities where mathematics is socially constructed, and where the goal is to establish the discontinuities between mathematics within the school and out-of-school. Such an approach is also useful when the goal is to integrate the complex interactions among affective, cognitive and cultural aspects. All the reasons stated above, together with our personal experiences of previous research, convinced the team that the most adequate approach for the goals of the study would be a qualitative and interpretative approach.

Therefore our research is also framed within the interpretative paradigm. We understand the educational reality as something constructed by the individuals involved in it at different levels of implication. It is our goal to come to know and to interpret the meanings attached by the different individuals involved to the situations taking place within the mathematics classroom. Moreover, the transition processes are essentially complex and, therefore, we cannot restrict our study just to a few variables, because by doing that, we risk losing comprehension in the understanding of those processes. Therefore we need an approach that integrates, as much as possible, different dimensions: the student as an individual, as a member of the classroom community, and as a member of a bigger community that provides him/her with a particular sociocultural identity (Lipiansky, 1990).

From this assumption, there are three standpoints that are useful to us:

a) The student as an individual. The students act towards people and their environment on the basis of the meanings they attach to these elements. As a consequence the meanings guide the actions and, to understand the different behaviours within the mathematics classroom it is crucial to understand the meanings underlying them.
b) The student as a member of the classroom community. The meanings are social products arising from and during the interactions between the different members of the classroom community. Thus, from a sociological point of view, we are interested in the extent to which meanings are shared or not.
c) The student as an individual with a sociocultural identity. The students attach meanings to situations, to actions, to themselves and to other people through an interpretative process, which is constantly revised and controlled through the acquisition of new experiences. Valorisation as a part of the interpreting process projects the sociocultural identity of the student within the mathematics classroom.

The focus of the qualitative methods used in the research is to establish connections between the individuals, their immediate context, that is the mathematics classroom, and the broader culture in which they live, both the new and the original. Taking into account this framework, we try to find ways for the individuals to make explicit their expectations, beliefs, values, and general understanding of situations, in order to interpret and give meaning to their actions and interactions. In our methodological design, and for a holistic comprehension, we focus on the understanding of the social phenomena as it is lived by all the individuals involved, and on the understanding of their perceptions of what is relevant and what is superficial. Therefore, different perspectives become relevant, from the students' to the teachers' contributions, including the researchers' starting points and assumptions. We understand that the research procedures used contribute to shedding light on all these perspectives with none being excluded.

Within the interpretative paradigm, the research procedures being followed by the team, are similar to those used by Abreu (1995) and Presmeg (1997) and, in relation to our different goals, consist of the following:

- interviewing teachers to find more about their understanding and beliefs of cultural conflicts in mathematics classrooms,
- interviewing students to investigate mathematical possibilities in their environment, and their expectations and valorisations of their learning mathematics,

- documenting and analysing examples of classroom situations and incidents that exemplify cultural conflicts, and
- creating and analysing classroom activities which are potentially rich, both from the point of view of research and of learning, because they could lead the students to project their knowledge, values and beliefs onto the activities, and thus allow either researchers or teachers to observe them.

4. CULTURAL CONFLICTS IN THE TRANSITION PROCESS: AS UNDERSTOOD BY TEACHERS

Too often, even if educational administrators, curriculum developers, principals and teachers accept the existence of cultural differences among the students as a reality, they appear to either reduce it to trivial facts, or in some way misinterpret them. Since the beginning of the project, we were conscious that the actual teaching of mathematics in multicultural classrooms does not greatly help the immigrant students' transition processes. As the teacher is the principal agent for change, despite the constraints from the superior structures and regulations that control the teachers' possibilities, we considered it to be of great importance to know more about teachers' understanding of cultural conflicts and differences among their students.

The research team considers that to promote any change in the teaching of mathematics that could help the co-construction of the students' transitions processes, we should first analyse how teachers understand and live the cultural conflicts arising in the social dynamics of the classroom. We accept that there is an inevitable aspect of the cultural conflict due to the fact that the classroom, as a part of an institution, institutionalises the students so that they have to give up part of their cultural and social identities. However, on the other hand, there is an aspect of the cultural conflict that could be minimised, the one that is related to the tensions between the culture which is familiar to the student and the school culture. Besides that, there is another part of the cultural conflict that should be avoided, and could be avoided, by considering the cultural nature of mathematics, and by the legitimisation of diverse forms of mathematical knowledge, not only the one represented by 'western' mathematics.

In our research we address these two aspects of cultural conflict that can be either minimised or avoided, and therefore we wanted to know more about the teachers' understanding of mathematics related to cultures different from the one established in our schools, and about their understanding of their students' transition processes. Moreover, we consider that to create coherence in the co-construction of the transition processes, to overcome the cultural conflict and to change it from a 'problem' into a potential source of richness, the first step is to be aware of the existence of the cultural difference. Therefore, in our research we analyse the understanding of the cultural situation by the people responsible for the teaching of mathematics.

During the first year of the project, the different members of the research team conducted several interviews with 20 teachers, both primary and secondary, men and women, who were working in schools with immigrant students. The schools where the interviewed teachers were working have different percentages of immigrant students (from 5% to 80%) and were located both in Barcelona and in small towns and rural areas. The teachers that were interviewed were the ones that after having heard a general description of the goals of the project were willing to participate and to help. Moreover, the teachers selected were all well known as 'good and experienced teachers' among the whole community of mathematics educators[5].

The interviews were semi-structured, with a standard set of questions and issues to be addressed but with the freedom to adapt the questions according to the reactions perceived and the answers given. Since the goals and assumptions were shared among the research team we accepted this freedom as a positive fact, enabling us to go deeper into the teachers' responses. The interviews had the aim to gain evidence about the teachers':

- expectations regarding the difficulties and potentialities of the immigrant students in their classes,
- opinions about how the students' social and familial contexts related to mathematics learning,
- acknowledgement of mathematics as a cultural product and how this was relevant to the learning of mathematics of immigrant students,
- actions to deal with cultural diversity in mathematics classrooms.

In the script for the interview some of the questions sought to reveal what were the teachers' opinions regarding the differences between the immigrant and the 'other' students. For instance: *Do you think mathematics is as useful for immigrant students as for the other students?* We stated the questions using intentionally the word 'other'. We did not want to establish definitions of the term that were 'politically correct' but to get answers that were relevant to our purposes. Some other questions sought to establish which were, according to the teachers, the more important barriers for the immigrant students to learn mathematics. Some of these questions were: *Which kind of student do you think can be successful? What are the characteristics of immigrant students that can lead them to be successful? What are the most important handicaps for them when learning mathematics?*

There were also questions addressed at knowing whether the teachers were aware of the non-uniqueness of mathematics and of the students learning mathematics in contexts different from the school one. They were questions such as: *Do you know if your immigrant students come from a rural or an urban context? Do you think that this fact can affect their learning of mathematics?* The questions connected with the social dynamics of the classroom, the interactions among the

[5] Our experience in the in-service teachers' education programs and within the FEEMCAT, the Association of Teachers of Mathematics in Catalonia, allowed us to contact what we considered 'good and experienced teachers'. Even if the selection of teachers under this definition can only be supported subjectively, we considered the group of teachers representative for our purposes.

students and with the teacher, and the relationship with the families had as a purpose to identify teachers' awareness of cultural conflicts in the transition process of immigrant students. They would also shed light on the possible actions they were developing in their classes to positivize the cultural conflict and to smooth the immigrant students' transition process. They were questions like: *Have you noticed if there is a particular way of organising the work in class that makes the immigrant students feel more at ease? Do you have a particular way of addressing them that can help their motivation? What would be your advice for a novice teacher who had to deal with a multicultural classroom?*

The following 'vignette' is part of the transcription of an interview conducted with a primary teacher, Maria, who has been working for 22 years and who in the last 12 years has been working in an school in a small town. In her experience as a teacher she has had about 25 students from Magreb. We have selected some parts of the transcript, to illustrate how rich the interviews were in shedding light on the important role of teachers acknowledging or not cultural differences among their students.

Interviewer: *Do you know if your immigrant students come from a rural or an urban context?*
Maria: *I have never thought about that before ... but, now I realise that this may be important for mathematics, and for other subjects....*
(... ...)
I: *Do you have any materials adapted to immigrant students?*
M: *I don't, not for them specifically. But I have some materials for the less able students and sometimes I use them for the immigrant ones.*
(... ...)
I: *Do the activities that the immigrant students develop outside school interfere with what they do in school?*
M: *There is a delicate moment, when they begin to go to the Mosque. They have a fear of going there to learn: there, to fail, not to know, is penalised, they are punished. When they begin going to the Mosque, they get closed!...Boys are punished, it is not the same with the girls, probably because it is not so important whether a girl learns or not.*
(... ...)
I: *What about the families? How do they see school? What are their expectations?*
M: *They come and ask 'does he do his duty'? They mean to learn things by memory, discipline and how they respect the teacher. They want to know if the teacher is happy with the child. They value the teacher's opinion. Their valuing the mathematics, or the relationship with the mates, is of second importance. They value the same the knowledge and the behaviour with the adults.*
(... ...)
I: *How do they respond when they are faced with the prospect of working in small groups? Do they adapt to it? When they can choose to work in small groups or individually, which do they prefer?*
M: *(no doubts in her voice). Individually: immigrant students, when they arrive into our country, they have to face the fact that other students interpret their not knowing the language and the habits as being stupid. 'He does not understand us, we say white and he does black!' That is why, when working, if the students can choose the mates in the groups, nobody wants to work with the immigrant ones, because 'he does not know'.*

From interviews like these we can conclude that teachers have different expectations regarding the potentialities of immigrant students, and different interpretations of the reasons for the difficulties the students face, and that they try to use different classroom organisations to face the multicultural fact in their classes. In general, through the different interviews done, we have evidence that mathematics teachers experience an enormous anxiety at having to face this new and complex educational situation, a task for which they do not feel prepared. They also feel they have very little knowledge of the transition processes that their immigrant students experience, and therefore they do not how to find the means to help the co-construction of those processes.

We want to present here the different kinds of teacher reactions that we have found in our research[6] to the challenge of having to deal with the multicultural groups in the mathematics classrooms:

- Those reactions of teachers for whom having immigrant students in their classes is a more difficult situation than they can handle and they try to ignore it. Through observing different classroom organisations in the schools we have visited, we have seen that some teachers even create classroom organisations in order to 'not see' the students belonging to ethnic minority groups, for example:

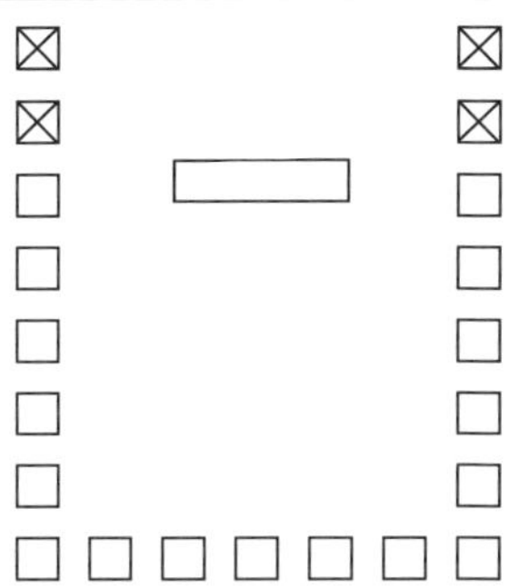

This shows the organisation of tables in a mathematics classroom. It is a class where the teacher usually faces the group having the blackboard behind him. The teacher's table is the one in the centre and the ones crossed are where the immigrant students sit.

- Those reactions of teachers who deny the cultural fact and state that doing any action related to issues concerning cultural differences or conflicts would generate an unequal treatment of the students, and would therefore be against equity. For example:

  In the past I have been very worried about how to teach a multicultural group of students, but now I am clear about it. (...) We have to normalise the situation of the students that come from abroad. One has to treat all the students the same, without pointing out any

[6] The following examples have already been presented as a contribution to the CIEAEM 51 conference, held at Chichester on July 99: Gorgorio, N., Planas, N, & Vilella, X. (2000c).

differences. By pointing out the differences it is much easier to fall into discriminating against them.

- Those reactions of teachers who reduce the cultural transition to a simple linguistic fact, and in no sense see mathematics as a cultural product:

  As soon as they learn our language, there are no more significant differences ... maybe in the social sciences, but certainly not in the mathematics classroom.

- Those reactions of teachers who feel the cultural diversity as a problem they are not prepared to face, and ask for ready made materials and recipes of successful methodologies:

  ... we do not know how to face these students ... their failure is really high ... our education through the teacher training programs is useless, it is too theoretical ... we need materials and methodological models to use.

There are data from our research that illustrate the fact that the ignorance, or the indifference, or the mathematics teachers' incapacity to deal with cultural difference, slows or inhibits the development of the potential that immigrant children have. The following example illustrates that fact.

Mohamed, an immigrant student, arrived into Catalonia and, when he was first at school, was diagnosed by his teacher as 'not knowing the basic mathematics algorithms'. He spent two years learning to add and beginning to subtract, not being able to communicate with the teacher. The following year, he moved to another town and his new teacher asked him if he already knew how to subtract when he came in. Mohamed, who already was able to understand and speak Catalan, was surprised at the question and showed his teacher 'his way' of doing 314 minus 182:

3	1	$\not{2}$	1	While adding: *I crossed out the 2, because*
1	8	3		*I have not practised for two years.*
4	2	2		

Nevertheless, we have also documented examples of teachers who develop teaching strategies, use learning materials, adapt the implemented curriculum and facilitate the immigrant students' transition processes through making explicit the norms that regulate the classroom social dynamics, thereby increasing the chances for the immigrant students to learn successfully. These examples of teachers that do a good job in helping the students' transition processes, together with the action-research developed in the schools where two of the members of the research team were working as teachers, convince us that the acceptance and the understanding of cultural difference within the mathematical class can contribute as a first step to overcoming the idea of the cultural conflict as a problem.

5. STUDENTS' TRANSITION FROM ONE CULTURE TO ANOTHER: MEANINGS ENDOWED TO THE NORMS

Mathematical enculturation (Bishop 1988) has been a key concept in research that studies classroom microcultures. We understand enculturation to be the process that inducts students into their familial culture. It is a learning process, through which the students acquire the culture of their group and, in particular, acquire the norms and the obligations of the community where they are living. From a sociological perspective the diversity of norms is a key element in understanding the phenomena taking place in the classroom (Whitson, 1997). Individuals develop their process of comprehension and of endowing meaning to facts when they participate in the negotiation of the norms, in particular, those that are specific to the mathematics classroom. To throw light on the norms and make explicit the beliefs about mathematical learning is closely linked with attaining the goal of students learning (Voigt, 1994, 1996). However, in some analyses of the students' learning processes, there is often the paradox of trying to explain them without previously documenting the norms that regulate the context where the learning takes place. The teachers' expectations regarding their students' behaviour and the students' expectations regarding their teachers' actions, which are the contributions of the ones accepted by the other and vice versa, and what is considered as 'correct' or 'acceptable', have all to be taken into account to understand the students' learning processes.

We consider, with Gravemeijer et al. (1990), that individuals adapt mutually to the actions of other members of the classroom community, and interact by reaching an agreement in the normative aspects and in the norms regulating the context. Several studies in the field of mathematics education (see, for instance, Yackel & Cobb, 1996, or Voigt, 1996) focus on the processes of establishing the basis of communication between teachers and students within the classroom. They analyse the negotiation to achieve shared meanings and the conflicts that appear as a result of the failure of that negotiation. In particular, they point out the relevance of sharing the meanings attached to normative issues for making communication possible. Nevertheless, knowing or accepting the norms is not a sufficient condition for communication. It is also necessary to be able to use them and to understand the connections among them, as well as the other demands coming from the learning situation.

The transition processes experienced by immigrant students can be studied through the analysis of the different meanings that the individuals participating in the classroom attach to the different moments in the classroom dynamics. Immigrant students have to adapt to new social and cultural situations, in particular to the social and cultural aspects of the classroom context, and they need to be helped to understand the meanings of the new situation and to reinterpret them according to their experiences, if we want them to be able to construct new meanings. The new situation that the students have to understand and reinterpret is plural, including, for instance, the different roles of teachers and textbooks (social norms), the different ways of understanding the learning of mathematics (sociomathematical norms) or the different uses of heuristics and algorithms (norms of the classroom mathematical practice).

To know more about the co-construction of meanings, during the academic year 1997–98, an important part of our research was developed in two public secondary schools (ages 12–16, compulsory school). The approach was action-research, and was developed in the IES (compulsory secondary school) M.T. in Barcelona, with about 90% immigrant students, and in the IES V., in the town of V., with about 5% immigrant students. While the first school was what we understand as a ghetto school, the second one is the only public school in the town, and the students attending it belong to all social classes. Núria Planas was teaching in the IES M.T. one group of 3rd year compulsory secondary students, 14–15 years old from 8 different countries. Xavier Vilella was teaching in the IES V. one group of 3rd year compulsory school students, with the majority of students being local but with a minority from Morocco. Both teachers were experimenting in their classes with learning activities that were designed by the research team with the expected goal of facilitating the students' learning and transition processes. They were using methodologies and teaching strategies that were considered respectful of cultural diversity and facilitative of communication within the class and that included, among others, open discussion, valuing each other's opinions and accepting out-of-school experiences. Moreover, in the first school, the teacher also conducted individual interviews with her students to know more about their expectations, values and self-images attached to mathematics learning.

The results we obtained are based on two essential aspects, triangulating methods for obtaining the data and triangulating perspectives in analysing them. The classes were video-recorded, the teacher was keeping a daily diary and there was an external observer who also took notes. The video-recordings were then seen by at least two members of the team, and the transcriptions were discussed within the team in our regular weekly meetings. In the discussions, we had the different perspectives of the four members of the group, enriching the process of analysis and giving credibility to the results.

Even if we consider that the interviews and the observation of the classroom experiences gave us enough rich data for our purposes, we want to point out that it was not an 'easy' task to work, and to do research in these schools. This was especially so in the first school, mainly due to its characteristics that make it a 'ghetto' school[7]. It has been difficult to access the classroom and to get the students to accept being interviewed. To study the development of the social dynamics we had planned to observe the class under its 'normal conditions', having one external observer sitting in one of the corners together with a technician with the video camera. To get the permission of principals and inspectors was nothing compared with the challenges we faced in the classrooms. These kinds of challenges have a direct impact on the possible research methods that could be used, and any of them needed to be strongly negotiated.

---

7 We consider the IES M. T. being what we have called a 'ghetto' school since, not only the percentage of immigrant students in this school was very high, but also because their population belongs to a highly deprived social and cultural class.

The first challenge was to have the students accept the presence of unknown persons within the classroom. The external observer had first to gain her credibility among the students by helping in the teaching of some sessions, before being allowed to be in the class observing silently and taking notes. The presence of the video camera within the classroom was also the result of a long and hard negotiation with the students, and in some cases, the families. In that sense, we went through some critical situations and the negotiation had to continue until we were sure that all the students really accepted the 'intruders'. One month elapsed before they accepted the video-recording of the lessons, provided there were no 'close-up' images of any of them. We are convinced that their acceptance was in terms of 'facilitating' their teacher's experiment, since they were told that she was working on a study to try to improve her teaching. Once again, this confirms the crucial role of the emotional aspects in the relationships established with the students when teaching in contexts with a social risk.

Concerning the interviews with the students, the only possibility was that they were conducted by their own teacher, regardless of any reluctance we could have from a theoretical perspective. This was only our first 'concession' during the negotiation process to gain the interviews. Some students did not agree to be interviewed at all but for those who did we also needed permission from the family or their legal tutors. In some cases we only got the permission under the condition of having some member of the family or from the community being present in the interview, a fact that certainly conditioned the students' answers.

In both cases, through the interviews or through their regular teaching, the teacher-researchers were heavily involved with the object of their study. The discussions within the research group of the data recorded and of the teachers' diary, together with collecting data in two different schools, all played a crucial role in compensating for the lack of distance between the two researchers and the phenomena that they observed as participants. The analysing of the data in the light of the existent theory took place within the working group and contributed to interpreting the situation, to making suggestions for implementing changes and to making theoretical contributions.

After the first phase of the project's work, we can claim that the different interpretations of the social dynamics of the mathematics classroom, and the dynamics of the mathematical activity among its members, are features that can interfere significantly with the actual teaching and learning process, and with the students' transition processes. To understand more about the normative elements of actions and interactions within the mathematics classroom, we shall adopt the constructs of social norm, norms of the classroom mathematical practice and sociomathematical norms, reinterpreting Voigt (1996), Yackel and Cobb (1996) and Presmeg (1998) as follows:

- ***social norms***, being the whole of the implicit and explicit norms that document the participants' structure within the classroom, and the dynamics between the teacher and the students, in the development of actions and interactions that take place in the class. Among them, we include the norms that regulate the organisation of the work within the classroom, the use of materials and learning aids, the discipline, and the use of talking time.

- ***norms of the classroom's mathematical practice,*** being the whole of the implicit and explicit norms that regulate the different possible mathematical behaviours of the teacher and the students. Among them we consider the norms that legitimate different mathematical strategies, processes and knowledge, or the different possible solutions to a particular task.
- ***sociomathematical norms***, being the whole of the implicit and explicit norms within the mathematics classroom, resulting from the juxtaposition of the social norms and the norms of the mathematical practice together with individuals' values, expectations, emotions, attitudes and beliefs. Among them are those that establish who 'has' the knowledge in class, or who regulates the valorisation of various forms of mathematics different from the 'official' one.

These are constructs integrated in the classroom's microculture and all of them refer to certain regularities present in the social interaction within the mathematics classroom that are established by the individual and group interpretation of what is perceived as acceptable or correct. We understand with Yackel and Cobb (op. cit.) that these constructs are useful to clarify the analysis of classroom situations and interactions, since they consider the possibility for different mathematical thinking processes, different social acts of participating and specific and unattended mathematical practices.

The research we have done until now confirms that immigrant students experience cultural conflicts in their transition processes, and that in the mathematics classroom cultural conflicts and disruptions emerge from the different understanding of the meanings attached by the different participants to its normative elements. Throughout the research, we have documented the way that individuals attach different meanings to the different norms that regulate the dynamics of the mathematics class. The following examples illustrate different understandings of the social norms, the sociomathematical norms, and the norms of the mathematical practice, together with some of the conflicts that different students experience because of their living in two cultures.

5.1. Social Norms

Sajid and Aftab are two Pakistani twin boys, aged 15, coming from a rural area near Karachi. They arrived in Barcelona one year ago, but have spent one year not attending school.

In the following 'vignettes' we see how Sajid expects a role from his teacher different from the one he understands she is playing:

Teacher: *Sajid, are you not interested in what we are doing?*
Sajid: *Yes Miss, I am*
Teacher: *Why do not you work then?*
Sajid: *In my country ... beat me, then I study. Here nobody beats ...*

and how Aftab's idea of what a classroom and a lesson is, is far from what he understands his 'actual' classes and lessons to be in Barcelona:

Teacher: *Could you not be quiet?!*
Aftab: *You know Miss, in my village, the school has no walls and we all have our place with a mark in the sand*
Teacher: *You do not like to be so many hours within the class, do you?*
Aftab: *I do not understand why classes need a wall*

Moreover, through the interviews with Sajid and Aftab and the classroom observation, we see how two students with very close personal histories (since they are twins) can live their transition processes very differently and react very differently to the new reality they face. Sajid works hard in class, does not easily communicate with his mates, is always very silent and closed, and lives 'anxiously' to succeed. Aftab, is always 'happy', wants to make friends, and is not so worried about learning. Even if both of them work as street sellers after class hours, Sajid brings the money home while Aftab spends it on clothes. During the weekend, Sajid works in a factory while Aftab spends the money he has earned during the week in the same activities as other boys of his age. When asked during the interviews what they wanted to become in the future, Sajid said he wanted to be a mechanic, while Aftab said: '*I want to be 'smart''*. Through the talking we understood that what he wanted was to 'assimilate' himself to those of his friends who, in his opinion, had achieved social success.

It is not only the role of the teacher or the class that are interpreted differently by the immigrant students, but also the role of textbooks, as shows the following 'vignette':

(The teacher had given the students worksheets for the session)
Nashoua: *May I bring the books I had in Morocco?*
Teacher: *To show them to me?*
Nashoua: *No! To use them*
Teacher: *What do you need them for?*
Nashoua: *To work with them, to know what the class is about!*

Nashoua is a 15 year old Berber girl from Morocco, who speaks Tarifit, the language of the Rif, a region in the north east of Morocco, and she attended school regularly in her country. She comes from a school culture where the 'book' is always present in class, and where the teacher is just an interpreter of the book. She feels 'anxiously surprised' since she has no book for most of the subjects she is following. She needs the book to feel at ease, to the point that she asserts that she does not know which class she is attending or which specific content she is studying unless she has a book.

Nashoua's transition process is not an easy one. She lives what, to an external observer, could be considered as 'contradictions'. As most of the Muslim girls of her age she follows Ramadan, the abstinence period prescribed by Muslim religion, and usually comes to class with a shaddor. However, at the end of the spring semester, while summer was approaching, she fainted in class. Her teacher was puzzled because Ramadan had been in January that year, so she asked her for the reasons for her sickness. She answered: *I fainted because I feel weak, I am on a diet, because I have to be thin for my new bikini to go to the beach with my friends.* Once more, we see one student that lives her transition processes full of conflicts when trying to reconcile the culture at home with the culture of her friends.

5.2. *Sociomathematical Norms*

We also have data that show that sociomathematical norms are differently understood among the different people participating in the same class dynamics. The following 'vignette' shows an example of how meanings and values associated with mathematics in itself – meanings and values that very much condition students' motivation and interest in learning – are also differently understood by the different participants in the mathematics classroom.

Teacher: *I want you to think, for tomorrow, of a mathematical problem or situation that can be linked with this photograph* (of a rural market with a woman selling)
Miguel: (the next day) *This was a trick! There is no mathematics problem, the woman has never been to school, she does not know mathematics.*

Miguel is a 16 years old gypsy student, who works cleaning houses, and who helps his family by selling in the weekend street markets. He has a very low opinion of his group regarding mathematical knowledge, and he is sure that if his people knew mathematics they would not be selling in street markets. He does not accept that there is a need to know mathematics to sell in a market and, therefore, he can not see any mathematics at all in this practice of his community. He only accepts as mathematics the officially established nature of mathematical knowledge within the school.

It is not only the idea of mathematics that is differently understood and valued among the participants, but also the idea of learning mathematics. For instance, Saima, an Indian girl, 15 years old, values and understands the learning of mathematics in a different way from the other students in her class, and the meaning she endows to certain situations in class are unexpected to her teacher and to the observers:

Saima: *Miss, I'm wrong in your class*
Teacher: *What do you mean? Why do you say that?*
Saima: *I do the same mathematics as boys, but I will not do the same work ... I do not want not be a mechanic. Please, can I do mathematics for girls?*

Through the interviews and the observation we could gain more understanding of Saima's transition process. Saima is the oldest daughter in her family, an extremely clever and beautiful girl, and always dressed in traditional Indian clothes. She participates in the class, and has several friends at school. She enjoys being with girls as with boys, but as soon as she leaves the school building and goes outside, she dismisses the boys, who do not understand why, even if she has tried to explain. She has been through a strong negotiation process with her family to get their permission to attend school and she knows that if her brother, who comes everyday to wait for her after school, sees her talking with a boy, this would mean the end of school for her. But she wants to become a teacher: '*like you miss, but only a girls' teacher*'. As far as we know, she lives happily in the two contexts, home and school, but she finds it difficult to establish connections between them.

The expectations of the mathematics teacher, as the person having the mathematical knowledge and the commitment to transmit it to the students, are also differently accepted or understood by immigrant students. Sheraz, a Pakistani boy, 15 years old, has difficulty accepting the role that he understands his mathematics teacher is playing. At the end of one class session, and since the students had not finished one of the activities, the teacher asks them to finish it at home. The students beg for the answer, but the teacher wants them to work more on it. The students insist, and the teacher, to avoid giving them an answer, 'excuses' herself by saying: '*I have not finished it either*'. Once the session is over, Sheraz, approaches the teacher, he is very nervous, even trembling, and shows his finished activity to her with not a single word, since his Spanish and Catalan are poor. The following dialogue illustrates a conflict in Sheraz's expectations about his teacher's knowledge:

Teacher: *I see, you have been able to finish the activity.*
Sheraz: (in a broken language) *Very easy.*
Teacher: (in a hurry, she is awaited in another room) *You will explain it to the whole group next day, will you?*
Sheraz: *I do not explain. Very easy.*
Teacher: (slowing down) *It is easy for you Sheraz, but may be, your mates or myself, we do not find it easy.*
Sheraz: (with a disrespectful tone): *What a teacher of mathematics you* (are)*! Better* (stay) *at home! Here students know more* (than) *teacher!*

Sheraz interprets his teacher's behaviour according to his experience, and he assumes that since she does not give the answer, she does not know it! This illustrates just one of the situations that convinced us that Sheraz does not accept a woman as a mathematics teacher, whom he considers as permissive, inadequate, and incapable. Moreover, he feels uneasy when addressing his teacher, he has a conflict between the respect due to her as a teacher and the feeling of frustration because she does not fulfil his expectations.

Sheraz is the son of a family that has left a position of high social class in Pakistan, and he lives his transition process with anxiety. He feels he is losing possibilities and wasting his time here. He expects to go back to Pakistan, therefore his efforts to learn Catalan or Spanish are not great. He does not accept his mates, even the others coming from Pakistan; in one of the interviews, when the teacher said: '*I see, you are from Karachi like Aftab and Sajid*', he answered very crossly: '*I am here because war, they are here because poor*'.

5.3. *Norms of the Classroom Mathematical Practice*

The data from our research also show that the norms of the classroom mathematical practice are differently understood by the different participants. One of the activities that we have used to document those differences is an adaptation of a well-known problem:

A farmer has 3 sons. When he dies he leaves his 17 cows to his sons. The oldest one must receive1/2 of the cows, the second 1/3 and the third, who is the youngest one, 1/9. How many cows will each of them receive?

The following 'vignette', from the transcript of one of the lessons in the IES V., corresponds to the whole classroom discussion after working in the problem in small working groups, and illustrates different ways of understanding the norms of the mathematical practice.

While some students are happy after 'solving' the problem within the context, and deciding that there is an error in the statement of the problem, some others seek a mathematical reason for the disagreement of the statement and the solution:

Carla: *The farmer was wrong, made an error*
Teacher: *An error?*
Lena: *The cows, they are 17, and this way it does not work*
Martí: *The cows are the ones they are. No, this is not the error, it does not work, but this is not the error.*
Teacher: *Where is the error, then?*
(… …)
Lena: *The error is in the fractions*
Carla: *What do you mean?*
Lena: *That the oldest one gets 1/2, the second 1/3 and the third 1/9 … this is not exact*
(… …)
Lena: *But 1/2 and 1/3 and 1/9 is not the whole amount, they do not add up to 1!!!*

Relating to the norms of the classroom mathematical practice, one of the facts that has proved to be more relevant is the role of the context evoked or the situation in which the mathematical problem or activity is embedded. The following example, shows once again, the importance of the context of a problem, as a tool to facilitate not only the involvement of the students in the problem by its appropriation, but also the appearance of other forms of solving processes that belong to what could be considered non-formal mathematics.

In the IES M. T., at the end of one of the sessions related to proportionality, the teacher asked the students to bring a recipe to work on in the next session. It was not a surprise either for the teacher or for the observer, that only one of the students brought a recipe the next day. By that time, both were conscious that the students would not consider it 'serious' enough for a mathematics lesson to work on cooking recipes! Just one of the students, Nadia, brought a recipe for a meat pie. The recipe was for 6 persons, and to make the pie, among other ingredients, 250 grams of meat were needed. The problem was to find out the appropriate quantities of ingredients to make the pie for 11 people.

The following 'vignette' illustrates the discussion about the quantity of meat required, after the students had thought about the problem individually and had discussed it in small groups:

Teacher: *Who wants to begin? Do we know how much meat we have to buy?*
Nadia: (raises her hand) *May I go to the blackboard?* (she goes, and writes) *458'333333....*
Teacher: *grams of meat?*
Nadia: *Shall I put the other ingredients?*
Teacher: *Wait, let us finish with the meat. Are we going to buy 458'333333.... grams of meat?*
Joel: (shouting disgustedly) *She is crazy!*
(Nadia erases the 3's and writes $458'\widehat{3}$)
Joel: *And what is that 'thing' over the 3?*
Nadia: *You shut up!*
Teacher: *Wait Nadia. Let us hear what Joel wants to say. Joel, good manners, please. Could you please tell us what's the matter?*
Joel: *She has never been shopping! We buy 500 grams, and everybody eats a little more!*
Nadia: *But you are inventing a new problem, it is for 11 people, not for 12!*

Nadia, a 15 years old girl from Morocco is probably the 'best' girl in the school in academic terms. She arrived last year, and had learnt Spanish very quickly. She is known among her mates as 'being clever' and always getting the 'right' answers. Nadia has always had 'good marks' in mathematics. At the beginning of the discussion of the problem, Joel, a 14 year old boy from Puerto Rico, said he could not work on that problem since he was on a diet and he explicitly decided not to work. However, when seeing what he thought was Nadia's 'nonsense' he could not refrain from joining in the discussion. Nadia, well 'trained' in formal mathematics had no problems in using symbols but Joel wanted meaning for the solution!

The two previous examples, together with many others we have collected during our research, confirm our idea about the importance of balancing symbols with meanings, and rigour with efficacy, and of legitimating the solving processes arising from contextualized problems, if the final aim is to acculturate the students into formal mathematics. However, we have identified some cases where even though the situation evoked facilitated the appearance of non-formal mathematics, it was also the situation in itself, probably due to the strong meaning attached to it, that prevented some students moving further towards a more formal mathematical strategy or reasoning. We are convinced that more research is needed to know at which point can the informal content of the lessons become an effective foundation for more formal mathematical reasoning and what are the limits of engaging students in contextualized problems.

The research done until now has also provided evidence of the existence of differences not only among the concrete mathematical knowledge that students have, but also among basic cognitive skills, confirming the findings of previous studies (see, for instance Cobb, 1989). We have found little evidence of differences among the first, probably due to the imposition of western mathematics onto other cultures (Bishop, 1990). In contrast, the differences among basic cognitive skills (intuition, abstraction, generalisation, deduction, contextualization...) seemed to be significant.

If one accepts the relationship between, on the one hand culture and mathematics, and, on the other, culture and cognition, one has to accept the possibility of identifying different mathematical potentialities and strategies related to the different cultures, as Abreu (1995) and Saxe (1991) state.

To end this section on the different meanings the participants in the classroom give to the different normative elements, we want to point out that throughout the development of the project, examples and situations like those of Sajid, Aftab and Sheraz, all coming from Pakistan, confirm for us that cultural transitions are something experienced individually. Even if they are shaped by the family, or the group culture, they are the result of a personal process, where emotions, affections, values and experiences play a bigger role than that of the broader culture of a country. It is important to stress that the different understanding of the norms within the mathematics classroom is a fact not only related to ethnic differences but also, and mainly, to individual differences closely linked to the experience every individual has or had.

6. SOME CONCLUSIONS AND SOME ISSUES TO BE DISCUSSED FURTHER

We understand research as a process whereby the perspectives of the participants are changed by its development. From that point of view, the progress of the perspectives of our own research group and its members has been really significant, and we believe that what we have learnt through its development could be helpful to other teachers and teacher educators. In particular, we have observed that students with an irregular school history, often belonging to cultural minority groups, also have their potential and capacity for creating strategies to face mathematical situations, which they show provided they are allowed to do so. However, often these students have interiorized negative social beliefs about the knowledge of their own social groups, and hide any mathematical knowledge that is different from that officially established in class. The traditional teaching of mathematics in western societies that considers mathematics as culture and value free, perpetuates cultural conflicts just by ignoring them.

We strongly believe that if we want any educational act to be positive both for the individuals and their communities, it would be helpful to begin to consider cultural differences as a source of richness, rather than of problems, in the educational context. However, it is an initial condition for any change that teachers become aware of, and acknowledge, the cultural differences among their students. From our research, through our experiences in in-service teacher education programs, and as members of associations of teachers of mathematics, we are convinced that the majority of teachers in our community are still a long way from considering that their immigrant students' failure, misbehaviour, lack of interest, or of motivation cannot be interpreted only through a simple lens of cognitive deficit.

In our educational context, there is still much to be done to help teachers understand that immigrant students live a complex transition process between cultures, with aspects like meanings, values and emotions being of equal, if not greater, importance than language. Therefore, there is a huge and urgent need to promote in-service teachers' courses, working groups and collaborative research teams, with the aim of helping teachers to design and implement learning activities and teaching methodologies that help the 'differences' to be made explicit, and to use them as starting points for building a rich teaching and learning resources. It is only through working from the realities of their classes that teachers will accept the possibilities for improving their immigrant students' learning processes.

However, it is not enough that in-service teachers' programs are aimed at developing materials or teaching strategies, they should also contribute to rethinking the whole meaning of mathematics education, as Abreu, Bishop and Presmeg state in the first chapter. In particular, the mathematics curriculum should be reconsidered. It is our judgement, that an approach to the curriculum that respects the idea of mathematics being a cultural product, and that implements it through 'humanising' it (Borba & Skovmose, 1997) could be a helpful one for facilitating the immigrant students' transition processes.

Even if the mathematics curriculum is defined at the pedagogical level, and essentially reflects ideas of mathematical content, we agree with Bishop (1988) when saying that the social and cultural structures establish a frame for this pedagogical development, especially concerning the values and ideology that the curriculum transmits. Curricular content implicitly hides and makes invisible the identities of the cultural groups that are different from the dominant culture, suggesting instead, a monolithic identity that is closed to cultural dialogue. Often it is the hidden curriculum that projects this 'uniforming' trend. The language used by the teacher, the attitudes adopted, the explicit values and behaviours, are an important part of the mathematics classroom's culture which is transmitted, even if unconsciously, to the students and which therefore in the end becomes part of the learned curriculum. Therefore, the first stage for making visible the cultural conflict is likely to be making the differences explicit to everybody in the teaching/learning process.

We are aware that creating cultural compatibility within the mathematics classroom is not an easy task for the teacher. However, this compatibility is a necessary step to changing the 'problem' of cultural difference into a rich resource. Thus, we would suggest an approach to the curriculum which is non-reductionist in its contents, articulating different meanings for every mathematical idea, and participatory in its methodology, to promote the contribution of all the students. The framework for such a curriculum can be built up from the six universal mathematical activities stated by Bishop (op.cit.), which allow us not only to establish connections between the mathematics we know and those of other cultures, but also to give plural meanings to mathematical ideas.

To articulate a participatory methodology, we would suggest a problem solving ambiance, which allows the establishing of connections between the different activities, and the introduction of mathematical situations that are real and significant to students with very different backgrounds. Problem solving can be a resource for treating the different curricular contents with authenticity, facilitating the contributions of all the students by opening the discussions, and accepting those linked with the particular cultural context of all the students. There is a clear difference in status between school mathematics knowledge and non-institutionalised mathematical knowledge resulting from the joint experience of a group (Fashesh, 1997). The school mathematics knowledge is embedded in the institutionalised knowledge, and its practices do not need to be justified since they are the result of the legitimated product of a community. Therefore, when students bring non-school knowledge into the mathematics classroom it is the role of the teacher to legitimate it and to facilitate the process of transforming it into institutionalised knowledge. Focussing the classroom dynamics on the discussion and protection of divergent points of view can help to legitimate the ideas linked to cultural contexts.

However, after three years working, we are aware that there are still plenty of unresolved issues that are of concern not just to us, but we suspect to anyone involved in such kind of research. For example, up to what point should the research in mathematics education reflect the real needs of the society in which it takes place? And how are these needs established? Who in charge is to decide which changes should be implemented in the mathematics curriculum in order to reach mathematics learning for ALL? Why, so often, do the bureaucrats of the educational administration pay so little attention to what is being done in the research field? Is it the fault of the researchers themselves because of not succeeding in communicating their ideas and results in a way that is useful for the classroom? Or is it because the research questions addressed are far away from the reality of the classroom? Why do researchers prefer working in contexts where there is less political and social commitment? Is it because the educational administration and the university system do not promote research that is really addressed to satisfying social needs?

In our particular situation, we would like to have, from the educational administration and the university, some answers, even if they do not fully satisfy us, to questions such as: Who is responsible for the growth and development of the project? Ourselves, as the researchers of course, but who else? And what is understood by the word 'researchers'? What are the respective responsibilities of the administration and the university which support the project, to facilitate the carrying out of the research? Who is responsible for any implementation of the ideas for the classroom and the schools? The teachers certainly, but not as isolated individuals. And then who else?

REFERENCES

Abreu, G. de (1993). *The relationship between home and school mathematics in a farming community in rural Brazil.* Unpublished Doctoral Dissertation, University of Cambridge, Cambridge – UK.

Abreu, G. de (1995). A matemática na vida versus na escola: uma questão de cognição situada ou de identidades sociais? *Psicologia: Teoria e Pesquisa*, 11(2), 85–93.

Apple, M.W. (1998). Markets and standards: the politics of education in a conservative age. In A. Olivier & K. Newstead (Eds.), *Proceedings of the 22nd Conference of the International Group for the Psychology of Mathematics Education,* (vol 1, pp. 19–32). Stellenbosch, South Africa: University of Stellenbosch.

Bauersfeld, H., Krummheuer, G. & Voigt, J. (1988). Interactional theory of learning and teaching mathematics and related microethnographical studies. In *Proceedings of the 2nd TME Conference,* (pp. 174–188), Bielefeld Antwerpen.

Bishop, A.J. (1988). *Mathematical enculturation: a cultural perspective on mathematics education.* Dordrecth, Holland: Kluwer Academic Publishers.

Bishop, A.J. (1990). Western Mathematics, the secret weapon of cultural imperialism. *Race and Class,* 32, 2.

Bishop, A.J. (1994). Cultural conflicts in mathematics education: developing a research agenda. *For the Learning of Mathematics,* 14(2), 15–18.

Bishop, A.J. (1996). A researchers' code of practice? Paper presented to *Working Group 24 of the 8th International Congress on Mathematics Education,* Sevilla, Spain, August 1996.

Borba, M. (1990). Ethnomathematics and education. *For the Learning of Mathematics,* 10(1), 39–43.

Borba, M. & Skovsmose, O. (1997). The ideology of certainty in mathematics education. *For the Learning of Mathematics,* 17(3), 24–28.

Bruner, J. (1990). *Acts of meaning.* Cambridge, Massachussets: Harvard University Press.

Cobb, P. (1989). Experiential, cognitive and anthropological perspectives in mathematics education. *For the Learning of Mathematics,* 9(2), 2–11.

Crawford , K. (1986). *Simultaneous and successive processing and executive control: individual differences in academic achievement and problem solving in mathematics.* Unpublished Doctoral Dissertation, University of New England, Armidale NSW- USA

Dowling, P. (1998). *The sociology of mathematics education: mathematical myths/ pedagogical texts.* Falmer: London

Eisenhart, M.A. (1988). The ethnographic research tradition and mathematics education research. *Journal for Research in Mathematics Education,* 19, (2), 99–114

Fasheh, M. (1997). Mathematics, culture and authority. In A. Powell & M. Frankenstein (Eds.) *Ethnomathematics: challenging eurocentrism in mathematics education,* (pp. 273–290). Albany: State University of N.Y. Press.

Geertz, C. (1973). *The interpretations of cultures.* London: Hutchinson.

Ginsburg, H.P. & Allardice, B.S. (1984). Children's difficulties with school mathematics. In B. Rogoff & J. Lave (Eds.), *Everyday cognition: its development in social context,* (pp. 194–219). Cambridge, Massachussets: Harvard University Press.

Gorgorió, N. (1998). Starting a research project with immigrant students: constraints, possibilities, observations and challenges. In P. Gates and T. Cotton (Eds.) *Proceedings of the First International Mathematics Education and Society Conference (MEAS1)* (pp. 190–196). Notthingham: Centre for The Study of Mathematics Education, Nottingham University.

Gorgorió, N. & Planas, N. (2000 a). Researching multicultural classes: a collaborative approach. In J.F. Matos and M. Santos (Eds.) *Proceedings of the 2nd International Conference on Mathematics Education and Society* (pp. 265–274). Lisboa: CIE, Facultade de Ciências da Universidade de Lisboa.

Gorgorió, N. & Planas, N. (2000 b). Minority students adjusting mathematical meanings when not mastering the main language. In B. Barton (Ed.) *Communication and language in mathematics education* (pp. 51–64). Auckland, NZ: The University of Auckland.

Gorgorió, N., Planas, N. & Vilella, X. (2000c). The cultural conflict in the mathematics classroom: overcoming its 'invisibility'. In A. Ahmed, J.M. Kraemer and H. Williams (Eds.) *Cultural diversity in mathematics (Education): CIEAEM51* (pp. 179–185). Chichester: Horwood Publishing.

Gorgorió, N. & Planas, N. (in press). Teaching mathematics in multilingual classrooms. *Educational Studies in Mathematics.*

Gravemeijer, K. (1990). Context problems and realistic mathematics instruction. In K. Gravemeijer, M. van den Heuvel and L. Streefland (Eds.) *Context free production tests and geometry in realistic mathematics education,* (pp. 10–32). Utrecht, Netherlands: RGME.

Kao, G. & Tienda, M. (1998). Educational aspirations of minority youth. *American Journal of Education,* 106, 349–384.

Lerman, S. (Ed.) (1996). Sociocultural approaches to mathematics teaching and learning. *Educational Studies in Mathematics,* 31 (monograph).

Lipiansky, E.M. (1990). Introduction à la problémaique de l'identité. In C. Camilleri, J. Kastersztein, E.M. Lipiansky, H. Malewska-Peyre, I. Taboada-Leonetti and A. Vasquez, (Eds.), *Stratégies identitaires.* Paris: Presses Universitaires de France.

Nieto, S. (1999). *Affirming diversity. The sociopolitical context of multicultural education.* New York: Addison Wesley Longman.

Nunes, T., Schliemann, A. & Carraher, D. (1993). *Street mathematics and school mathematics.* Cambridge: Cambridge University Press.

Ogbu, J.U. & Simons, H.D. (1998). Voluntary and involuntary minorities: a cultural-ecological theory of school performance with some implications for education. *Anthropology & Education Quarterly,* 29(2), 155–188.

Presmeg, N.C. (1997). A semiotic framework for linking cultural practice and classroom mathematics. In J.A. Dossey, J.O. Swafford, M. Parmatie and A.E. Dossey (Eds.), *Proceedings of the XIX Annual Meeting of the North American Chapter of the International Group for the Psychology of Learning Mathematics,* (vol 1, pp. 151–156). Columbus, Ohio: ERIC SMEAC.

Presmeg, N.C. (1998). *Comparison of qualitative and quantitative research methods.* Paper presented at the TIEM98 seminar on Research Methods on Mathematics Education. Centre de Recerca Matemàtica, Institut d'Estudis Catalans, Barcelona, Spain.

Rasekoala, E. (1997). Ethnic minorities and achievement: the fog clears. *Multicultural Teaching,* 15(2), 23–29

Saxe, G.B. (1991). *Culture and cognitive development: studies in mathematical understanding.* Hillsdale, New Jersey: Lawrence Erlbaum Ass. Publishers.

Schoenfeld, A.H. (1994). A discourse on methods. *Journal for Research in Mathematics Education,* 25(6), 697–710.

Voigt, J. (1994). Negotiation of mathematical meaning and learning mathematics. In P. Cobb. (Ed), *Learning mathematics. Constructivist and interactionist theories of mathematical development* (pp. 21–50). Dordrecht, Netherlands: Kluwer Academic Publishers.

Voigt, J. (1996). Negotiation of mathematical meaning in classroom processes: social interaction and learning mathematics. In L.P. Steffe & P. Nesher (Eds.), *Theories of mathematical learning.* Hillsdale, New Jersey: Lawrence Erlbaum Associates.

Whitson, B. (1997). Language minority students and school participation. *Educational Researcher,* 26(2), 11–16

Wilson, P. & Mosquera, P. (1991). A challenge: culture inclusive research. In R. G. Underhill (Ed.), *Proceedings of the XIX annual Meeting of the North American Chapter of PME,* (vol 2, pp. 22–28). Blacksburg, Virginia.

Yackel, E. & Cobb, P. (1996). Sociomathematical norms, argumentation, and autonomy in mathematics. *Journal for Research in Mathematics Education,* 27(4), 458–477.

CHAPTER 3

THE TRANSITION EXPERIENCE OF IMMIGRANT SECONDARY SCHOOL STUDENTS: DILEMMAS AND DECISIONS

ALAN BISHOP
Monash University

1. INTRODUCTION

Being an immigrant school student in a new country is a difficult matter. Language problems predominate, compounded by not knowing which other students to trust in the school, not knowing the school rules (except that you know there are likely to be many school rules), and not knowing the teachers:

Interviewer: How good do you think your teacher thinks you are?
Tra: I don't know, no idea. It is hard to know what the teacher thinks.

The social pressures on the students are immense. As well as their own self-imposed pressures to survive in the new environment, there are pressures from their parents who may be ultimately dependent on their ability to earn money for the family. There are the usual learning and assessment pressures from their teachers which although well-meant, are not always well-received, and there might also be the less well-meaning pressures from some of their peers who may not be willing to accept them into the social milieu of the classroom or the school.

These social pressures are exacerbated by the cultural conflicts experienced by every immigrant person, but particularly by immigrant students. It is a particular problem for students because of the predominantly cultural nature of schooling. Education is a process of cultural induction, and schooling is the formal instrument of the induction process. As an immigrant bus driver or nurse, one might well be able to 'slip into' the same job in the new country, as many occupational practices are similar from country to country, although there too will be cultural differences causing conflicts and anxieties. Schooling, although superficially similar in different societies, differs markedly in its cultural framing.

Moreover, whilst for a bus driver or nurse the expectations of society for the newcomer will almost certainly be spelt out (if only for health and safety reasons) there is no such provision for the immigrant school learner. There will probably be some

G. de Abreu, A.J. Bishop, and N.C. Presmeg (eds.), Transitions Between Contexts of Mathematical Practices, 53–79.

descriptions of aspects such as the appropriate clothes to wear, the books and materials needed, the times of the school day, the sports provisions etc. No-one though describes how one is supposed to operate in the new learning environment, which other students to pay particular attention to in class, the new social niceties of how to make friends, or which rules are to be obeyed and which are negotiable. It is never written down which teachers are particularly sympathetic to the plight of the immigrant student, or which ones are to be avoided. There are no manuals explaining what counts as an acceptable reason to be late for class or not to have done the homework, or how much work needs to be done in class. These are all part of the hidden curriculum of the school and classroom and are necessarily never written down. But students who have ever been to school before know just how important the knowledge of such hidden 'rules' is.

This chapter is especially concerned with analysing aspects of the transition experience of immigrant secondary school students in their new mathematics classes. Learning mathematics in school classrooms is a particular kind of mathematical practice with its own criteria and rules. It is not simply a matter of learning mathematics with a teacher. There is a variety of special tasks, done to certain standards, according to certain procedural rules, and in a social context that is not well-defined partly because it is continually being reconstructed by the participants. For the newcomer there is a lot to learn, and perhaps the least of the problems is the mathematics itself.

A learner's mathematical practice is shaped and negotiated by the classroom participants, but not all participants have equal power in that shaping. The teachers have of course the power invested in their position, but that is only power of the formal, institutional kind. It is the kind of power that allows the teacher to decide on the implemented curriculum in the class, the kinds of activities to be done, the materials to be used etc. Effective power, the kind that enables the teacher to shape interactively a learner's mathematical practice in the classroom, must be earned and won through negotiation and respect.

The classmates or peers also play a fundamental role in shaping learners' practices, again not all equally. There are 'significant others' among the peers who will be particularly influential for the new student, and they may have particular attributes, such as age, ability or a particular personality. However studies of influence (e.g. Moscovici, 1976) suggest that it is as much the learner who choose whom to be influenced by as it is the significant others choosing whom to influence. The suggestion is that it is the learner who chooses which 'others' will be 'significant'.

The learner has in some sense the most power over their learning, choosing for example how much effort to expend, whom to listen to, and whose views to respect. But also they are in another sense a party to the power of others, being a product of their cultural and social history, a history shaped in large part by their family life and by their life outside school. Their parents in particular are likely to be influential, and in the case of mathematics learning it seems from the interviews quoted later that it is often the father who is the more influential of the parents.

Much of this is likely to be the case for all mathematics students. In the social situation of a mathematics classroom they must learn to 'read' the social dynamics, the social interactions, and the social and body 'language'. For the immigrant mathematics student the situation is much more challenging since all of this must be learnt afresh.

Language is a useful metaphor here. When one learns one's first language there is little awareness that one is learning a language. This usually occurs when one begins to learn a second language. That is when one becomes acutely aware of the phenomenon of language. The same is true for learning to 'read' a mathematics classroom. In one's home country and culture, learning to read one's classroom comes naturally, and one is rarely aware of doing it. Of course that does not mean that it is an easy matter, far from it. But in another country, and in another classroom, one quickly realises that one does not automatically understand the 'language' of that classroom. The signs and symbols must be learnt afresh. One is in a transition situation.

Language usage, literally, is of course the most obvious and pressing aspect for all students, as it is the source of all communication of meanings. For the teachers also the students' language use is a strong indicator of the effectiveness or otherwise of the acculturation process. Here is an example of the kind of challenge that the immigrant student faces in mathematics classrooms:

Interviewer: Do you ever do any maths in English?
Ty: Yes, in the Philippines, I lived there for 2 years before I came here.
Int: But you are originally from Vietnam?
Ty: Yes, Vietnam to Hong Kong and then the Philippines.
Int: It is a long way round?
Ty: Yes, for 7 years.
Int: It must have been very hard?
Ty: Yes.
Int: When you are doing maths in the classroom do you think in Vietnamese?
Ty: I think in English.
Int: Do you ever think in Vietnamese when you are doing that maths?
Ty: No never, the solution is different. We have 4–5 ways of solving it.

It is hard to avoid the conclusion that life in the mathematics classroom would be fairly complicated and challenging for that student, even though he admits to being able to think in English, which is already a tremendous advantage for an immigrant into an English-speaking country. However learning mathematics through one's second or even third language still presents a unique set of problems (Dawe, 1983) that can hinder one's mathematical development and progress. More positively however, it might be possible that out of these contrasting language and cultural experiences immigrant students could be constructing a richer understanding of mathematical ideas than we might otherwise expect.

This chapter explores some of the significant aspects of this transitional mathematical practice in order to see what light can be shed more generally on the sociocultural milieux of mathematics classrooms, and on how to improve the quality of mathematical teaching and learning for all students.

2. CONTEXTUAL BACKGROUND TO THE RESEARCH

Immigration into Australia has created an extraordinarily varied cultural history in the country. From the early days when the English, Scottish and Irish immigrants brought their mix of Anglo-Celtic culture with them, to today when there are estimated to be more than 150 languages spoken, not counting the Aboriginal languages, Australia has seen wave upon wave of new immigration from different parts of the world. Coming as refugees, or as members of families of existing Australian citizens, or as new migrants anxious to create a new life for themselves, the new immigrants have all brought with them their languages and their cultures. This mix of languages has therefore been changing constantly from the 1950s when the immigration policy for Australia meant a heavy influx into schools of Italian and Greek speaking children, through waves of other European groups to the mixture of Asian, African, and sub-Asian continent groups migrating at present.

The extent of the migration is shown by a report by the Public Affairs Section of the Department of Immigration and Multicultural Affairs, Canberra in 1996 which states: 'Today nearly one in four of Australia's 18.5 million people was born overseas. In 1995–96 the number of settlers totalled 99 139. They came from more than 150 countries. Most came from New Zealand (12.4%), the United Kingdom (11.4 %), China (11.3%), Hong Kong (4.4%), India (3.7%) and Vietnam (3.6%).' (Sources: Australian and Immigration 1788–1988 and other material produced by the Department of Immigration and Multicultural Affairs.)

Most Australian schools have a significant minority of non-English background students attending, and in their classrooms there can be a number of languages represented there. Although normally all teaching is in English, many classrooms can have up to 10 different languages spoken by the students, and some have more than that. Clearly the multicultural and multilingual situation in Australia creates special challenges for learners, teachers, families and policy-makers. It also presents ample opportunities for those concerned to research sociocultural aspects of mathematics learning and teaching, as a way of developing ideas for mathematics education in any multicultural society, not just Australia. Most countries in the world are experiencing rapid increases in migration, and school populations in many countries are becoming much more multicultural.

Earlier Australian research on cultural issues, summarised and reviewed by Atweh, Cooper and Kanes (1992), and by Ellerton and Clarkson (1995) in the context of language, generated a variety of literature. For example, Howard (1996) discusses his important research with Aboriginal teacher educators, making us aware of the importance of sensitivity in dealing with cultural issues. Leder, Rowley and Brew (1995c) present some challenging data about students' performances in the Victorian Certificate of Education, the end-of-school examination in the state of Victoria, which suggests that immigrant students often outperform their 'local' counterparts. Thomas (1995) argues for improving language policies in schools to improve the chances of Non-English Speaking Background students succeeding at University mathematics.

This chapter is based on a research project, funded by the Australian Research Council, carried out in Melbourne, Australia, over a three year period, and undertaken in collaboration with Gilah Leder, Chris Brew and Cath Pearn. The basis for the theoretical framework was the construct of 'cultural conflicts', first elaborated in Bishop (1994) and discussed briefly in chapter 8 of this book.

In this research project the focus was on secondary school students, though not on immigrant students specifically. However one of the main reasons for proposing the project was that the previous research on such students had produced conflicting findings. Some studies showed that immigrant students were achieving better than local students, while other studies showed the reverse to be true (for example, see Leder et al., 1995c). The present study was formulated in the hope that, by considering the students' situation from the cultural conflict perspective, we would develop some richer and more illuminating understandings of the immigrant students' transition experience in mathematics classrooms in Australia.

Research on the cultural conflicts experienced during transition in the learning of mathematics has played a minor role in the mathematics education scene in Australia. However one hope is that the work of the project will demonstrate the potential and the promise of this kind of research, not just to benefit immigrant and Aboriginal students, but the school population more widely. So the scarcity of such research could not be due to any lack of potential benefit nor of interest. Australia, with its significant Aboriginal population, also contains the world's second highest variety of migrant country 'backgrounds', and would seem therefore to be a research site of huge significance for this theme.

As well as having intrinsic and pragmatic interest in terms of the contribution such research could make to the development of the unique multicultural nature of Australian society, there also seemed to be a large potential pay-off for the development of richer theoretical constructs. It is only relatively recently that mathematics educators have taken seriously the cultural nature of mathematical knowledge, with all that that idea implies, and most research and development in mathematics education still privileges a mono-cultural, or even a non-cultural, image of the subject.

However one crucial reason for the scarcity of this research could well be the inherent challenges of doing it. It is indeed a complex research field. As will be shown in this chapter, and as is shown elsewhere in this book, not only is it complex in terms of its theoretical constructs, and in terms of the educational issues involved, but there are specific and huge complexities in terms of the practicalities of doing empirical research in the area. Thus as well as having great practical and theoretical importance in Australia and in many other countries, research on cultural conflicts in mathematics education faces researchers with many other challenges that, if they can be met, may well enable the generation of ideas of potential benefit for the whole field of mathematics education.

3. CHOOSING THE RESEARCH SAMPLE: WHO ARE THE IMMIGRANT STUDENTS?

It was initially hypothesised that the typical school mathematics culture in Australia could be characterised as being predominantly 'Anglo, male, and middle-class'. It was argued in the research proposal that the study should involve the variables of gender, ethnicity and Socio-Economic Status (SES), and that these three variables would assist in both the selection of the students, and the interpretation of students' experiences with any kinds of conflicts they were having with the typical Australian school mathematics culture. The chapter will draw on some of the same data that was first reported in earlier papers (e.g. Leder et al., 1995a and b; Bishop et al., 1996 and 1997, and Pearn et al., 1996), as well as presenting other analyses and other evidence from the study that focuses on the immigrant student experience.

The first issue to be faced concerned the process of selecting the students for the study, and it was decided to concentrate on year 7 and year 9 students (approximate ages 12 and 14 years). This is an important age group for teachers as their students become more aware and articulate about the mathematics education they are experiencing. The selection process necessarily involved using classifications and categorisations. So how does one describe and find immigrant students? And how does this process affect the research study?

Educational institutions are replete with bureaucratic procedures and one of these concerns categorising and labelling students. Whilst this may seem to be a necessary activity for the smooth administrative running of a complex institution such as a school, it also serves to emphasise certain differences among the students rather than others. Students are often categorised and then grouped in terms of such aspects as age, gender, mathematical ability, academic aspirations, language and cultural background, not to mention intellectual or physical disability. Categorising and labelling are effective forms of control, and are not so hidden parts of the hidden curriculum of the school and of the education system. Even in the formulation of the research project, in order to obtain government funding it was necessary to use the commonly accepted variables of gender, ethnicity and Socio-Economic Status together with their accepted classification systems. However, with the exception of gender, the use of these 'commonly accepted' variables was not unproblematic, nor were they particularly helpful from an interpretative point of view.

There are two kinds of governmental classification which relate to the educational provision for immigrant students in Australia, both of them based on language, arguably the most obvious difference criterion, and the aspect that appears most obviously in the teaching context. One category refers to 'English as a Second Language' (ESL) students and the other to 'Non-English Speaking Background' (NESB) students (I am grateful to my colleague Jill Brown for her clarification of these terms below). Students are categorised as ESL students if both the following conditions are met:

1. they have been living in Australia for not more than seven years prior to 1 Jan in the year in which they are studying Year 12

2. English has been their major language of instruction for a total of not more than seven years prior to 1 Jan in the year in which they are studying year 12.

If students meet the above criteria, a number of provisions have been made for them in VCE (Victorian Certificate of Education), which is the examination at the end of schooling in the state of Victoria. The most important provision is that they are able to study English-ESL, rather than English at Year 12 level. The course is similar enough to maintain comparable status for tertiary entry, and the main advantages are firstly that the ESL students compete against other ESL students rather than against native speakers and secondly that the examination is set and assessed by ESL specialists

Students are classified as NESB students if they were born in a non-English speaking country, or if they were born in Australia and one or more of their parents were born overseas in a non-English speaking country. Schools are allocated funding for special ESL staff on a complicated formula based on the numbers of students in the school, and the year of arrival (so a recently arrived student is 'worth more' in terms of staffing than one who has been in the country for longer). However NESB students born in Australia are 'not worth' anything financially to the school – even though their needs are still very real. Such is the controlling power of the labelling process.

After considering the available classifications and data it was decided to use schools' NESB intake as the criterion for selection as it offered a wider range of schools and students. Eight State co-educational secondary schools, mainly urban/suburban in their character, were selected with varying cultural 'mixes'. This then was the formal planned procedure.

However as the research progressed from school level analysis to individual student level analysis, it became increasingly apparent that the ESL/NESB classification was suspect, and unhelpful. While the ESL/NESB labels may help funding and examination decisions at the state and national levels, there are too many inconsistencies and overlaps for good research at the individual level. For example, there are confusions about how to classify Hong Kong students, as they are officially considered NESB students, even if their parents are of Anglo-cultural background, and Cook Islanders are not officially classified as NESB students even though their first language is not English. Often schools classified pragmatically a student as ESL because of his/her English language problems despite being here for more than seven years. Also some ESL-classified students could have received all their schooling in Australia and in English.

Accordingly, for individually-based data analyses, a new classification was created and the following criteria used for the ethnicity classification of individual students, referred to as Non-Anglo-Cultural-Background (NACB) students. These were students:

- who were born into a non-English speaking home, regardless of country,
- whose first language was not English and was the language predominantly used at home,

- whose observable characteristics and language usage in class (if other information was unavailable) suggested that they should be called NACB students.

The contrast being made was between the NACB students and the ACB students who were students born into an English-speaking family environment using English as their first language. One unfortunate aspect of this NACB labelling is that, like the NESB label, immigrant students are again negatively categorised – it says what the students are not, rather what they are.

A more general issue with institutional labelling however is that the official definitions that exist will tend to be used. As Mehan (1996) and McDermott (1996) point out, official definitions can trap students into a web of institutionalised arrangements that may well not be of benefit to the individual concerned. Indeed, the title of McDermott's influential paper 'The acquisition of a child by a learning disability' is an excellent case study of how an institutionalised construct gets applied to particular learners. The constructs existed before these particular students arrived at the school, and because they exist they will in general be applied.

It is only therefore a matter of which students get 'chosen' by the label. Moreover once an 'educational label' has been applied to a student, it is extremely difficult for anyone to remove it. This is not to say that there are no teaching or learning benefits accruing to the students so chosen, but only that there may also be inherent dangers in the use of such labels in contexts for which they were not intended, including interpretative research.

4. DATA COLLECTION PROCEDURES

Having discussed the process, and the problems, of selection of the students let us now turn to consider the processes of data-collection. This is an equally complex problem, particularly when researching cultural issues. The research team's interest in cultural conflicts, and its previous research in this area led it to focus on attitudinal, affective, and socially-related responses. In addition both group and individual data were sought. Group data were important, in order both to see if any general patterns of response differences and contrasts occurred and also to enable the 'location' of particular students within the spectrum of the research sample as a whole. Individual data were crucial if any progress was to made in understanding the immigrant student experience in the social environment of the mathematics classroom.

Data was therefore gained from four sources: a multi-dimensional questionnaire to all the students, video-tapes of three lessons in each class, which were used to observe the selected students, interviews with the students, and interviews with their parents where possible. Students' individual questionnaire responses would act as the initial context for discussion in the interviews.

The questionnaire was developed using some items from the Fennema-Sherman Attitude Scales (Fennema & Sherman, 1976), the Mathematics Attribution Scale (Fennema, Wolleat & Pedro, 1979), and the Individualised Classroom Environment Questionnaire (Fraser, 1990). Other items were developed by the researchers based on statements from the Australian mathematics curriculum document *The National Statement* (Australian Education Council, 1991). An important consideration in the development of the questionnaire was the need to ensure that the language used in the questionnaire was also appropriate for the administration to NACB students.

The questionnaire had four distinct sections. *Your Views about Mathematics* assessed students' perceptions of mathematics. *More Views about Mathematics* assessed students' attributions for success and failure in mathematics: in relation to ability, effort, task and environment. *How Good Are You?* assessed the students' perceptions of their own achievement level in mathematics and how they thought their parents, teachers, and peers would rate their mathematical achievement level. The *Individualised Classroom Environment* questionnaire was used to obtain each student's perception of the learning environment within the mathematics classroom. Most items required students to respond on a five point scale from Strongly Agree to Strongly Disagree. The questionnaires were in English, but they were administered personally by the research team who had ample opportunity to explain any difficulties to the students.

The most challenging part of the data gathering process however was gaining the interviews with the parents. These were either carried out at home, at the parents' work-place, or at Monash University, according to the parents' choice. There were particular difficulties with gaining interviews with the parents of the immigrant students. The ethical procedures rightly permit any chosen subjects to decline to participate in the research, and unfortunately several of the immigrant students' parents took that option. Whether this was due to language worries, being too busy, or for fear of giving information to strangers, is of course not known. Whatever the reason, the difficulties of engaging these parents is one of the problems to be overcome in doing this kind of research.

5. INFERENCES FROM THE GROUP DATA

The questionnaire produced some intriguing results that are summarised in the next 4 sections. The data presented here are in relation to the NACB/ACB individual student classification.

5.1. Students' Views About Mathematics

In this part of the questionnaire there was not a great deal of difference shown between the students, and Table 1 presents the significant results from the analysis done using the NACB/ACB classification described earlier:

The data show that the NACB students feel less anxiety and have greater confidence, have a preference for logical and 'one right answer' mathematics, and have a stronger view of mathematics as a male domain. Given that there has been much emphasis paid in Australian education to overcoming the prejudice about girls doing mathematics, we can perhaps see here a cultural difference, in that the NACB

Table 1

	NACB students			ACB students				
	mean	s.d.	n	mean	s.d.	n	F ratio	p
Anxiety	2.28	0.76	177	2.44	0.82	178	5.136	<.05
Confidence	3.79	0.79	179	3.63	0.91	178	4.022	<.05
Maths is not a male domain	4.10	0.64	175	4.34	0.57	179	20.21	<.001
Maths is logical	4.01	0.82	182	3.76	1.04	178	7.28	<.01
In maths there should always be one right answer	3.12	1.32	183	2.84	1.21	179	4.546	<.05

Scale: 1 = strongly disagree to 5 = strongly agree.

students appear to have brought with them a stronger cultural belief in the 'maleness' of the mathematical domain.

In the same way the 'logical' and 'one right answer' results could reflect the fact that in Australia there has been an emphasis in recent years on more open-ended mathematical activity, such as through investigations and projects. This does not appear to be the case in many of the 'home' countries of the immigrant students.

5.2. *Attributions of Success and Failure*

For many years attribution theory has been a powerful lens with which to view students' interpretation of their own achievements (see e.g. Parsons, 1981) and has been particularly helpful when analysing girls' mathematical performance. Data on attributions often produce interesting results, and those in this study were no exception. The questions offer respondents the choice between attributing successes and failures to four factors: ability, effort (the 'internal' factors), task and environment (the 'external' factors). While there were no significant differences in relation to the external factors, Table 2 shows three significant differences and one trend between the NACB and the ACB students in relation to the internal factors:

This pattern of results is interesting, as it shows that the NACB students rate their ability and effort to be significantly more important for their successes than the ACB students, while it is the reverse for their failures. The high emphasis given by the NACB students to the variable *effort* is particularly interesting. It may point to a cultural artifact, with for example students from Confucius-Heritage-Cultures (Leung, 1998) such as Chinese, Japanese and Vietnamese, often attributing their success to this aspect. Or it may point to a social aspect, such as the immigrant students' recog-

Table 2

	NACB students			ACB students				
	mean	s.d.	n	mean	s.d.	n	F ratio	p
Anxiety	2.28	0.76	177	2.44	0.82	178	5.136	<.05
Success due to ability	3.53	0.76	174	3.34	0.84	174	3.442	<.01
Success due to effort	3.64	0.77	176	3.48	0.79	175	3.94	0.077
Failure due to ability	2.90	0.78	172	3.13	0.84	175	2.433	<.01
Failure due to effort	3.06	0.78	173	3.26	0.72	174	5.426	<.01

nition of their need to expend greater effort in order to survive in the new teaching situation.

5.3. Students' Perceptions of Their Level of Achievement

This part of the data comes from a short but revealing questionnaire which asks the students to rate their own level of achievement, and what they perceive their teacher's, their peers' and their parents' opinions are about their level of achievement. The ratings go from 1 (weak) to 5 (excellent). Table 3 shows four areas of significant difference between the NACB and the ACB students:

Table 3

	NACB students			ACB students				
	mean	s.d.	n	mean	s.d.	n	F ratio	p
Self rating	3.51	0.92	189	3.27	0.96	180	3.314	<.01
Perceived teacher rating	3.50	0.98	186	3.19	1.01	178	5.811	<.01
Perceived class mate rating	3.66	1.02	186	3.22	1.11	179	8.950	<.001
Student wished-for rating minus self-rating	1.21	0.86	189	1.44	0.91	180	2.808	<.01

In each of the first three cases the NACB students rated themselves significantly higher than the ACB, which supports the earlier findings about their feelings of greater confidence and less anxiety. They also seemed more 'content' with their performance, in that their 'wished-for' rating was not as far ahead of their self-rating as it was for the ACB students.

Regarding the other influences on the students, there was an interesting finding regarding the gender of the students. The teachers were asked to rate their students' achievement levels and these were compared with the students' self-ratings of their achievement, although not all teachers agreed to do this. However the results showed that compared with the teachers' ratings, over half the students over-rated their achievements, with the over-rating being generally more prevalent among the boys and under-rating among the girls. An interesting contrast came with the comparison of the students' own ratings with their perceived classmates' ratings of their achievement level. In this case girls were more likely to indicate that their class-mates would over-rate them and boys were more likely to indicate that their peers would under-rate them.

5.4. Summary of the Group Data

Having imagined at the start of the project that the transitional mathematics learning experience for the NACB immigrant students in this study would essentially be one of difficulty caused by the conflict experience, it is rather surprising to see a much more positive picture emerging from the data. It is also now possible to speculate a little more confidently about the subtle, and sometimes not-so-subtle, ways the transition conflicts play out.

On the one hand there seems to be an interestingly coherent set of perceptions and attitudes of the NACB students, when compared to the ACB students. They generally appear to be more confident, less anxious, and more at ease with their perception of their own abilities. They generally attribute their successes to their own abilities and efforts, but not their failures, and they feel that their teachers and their classmates rate them higher than do the ACB students. Furthermore, NACB students perceive that their classmates rate them at a higher level compared to the rating they give to themselves.

On the other hand, as can also be seen, there are different possible interpretations of these general conclusions. For example, the final conclusions above may not necessarily be good news. The higher rating from teachers and classmates, as perceived by the students themselves, may well increase the pressures on them to achieve at a higher level than they would like. Also they appear to have a preference for logical and 'one right answer' mathematics, and may well be less than comfortable with teaching approaches that emphasise discovery, investigations, and the use of intuition. They have a stronger view of mathematics as a male domain, and once again may not enjoy teaching approaches that include such features as group work and discussions, designed in part to encourage a greater participation of girls.

Having looked at the important group data coming from this study, and drawn some interesting conclusions, it is necessary to move now to the interview data. Whatever the relative merits or otherwise of classifications such as NESB, ESL or NACB for collecting and analysing group-based data, there is always a need to progress to the individual level to enrich our interpretations. This is particularly so in this case. There is no reason to think that the immigrant transition experience is the same for all students, and individual circumstances will always affect the transition. Having said that, it will be important to see if there are any general aspects that emerge from the interviews.

6. INDIVIDUAL STUDENTS IN TRANSITION

6.1. Classroom Interactions with the Teacher

The first aspect concerns how the teacher and the immigrant student relate to each other. This is of course one of the most important features of the transition experience for the students, and reveals much about their feelings. The extracts also add some specific nuances to the bald data above.

For example, consider these two extracts from the interview with a Year 9 immigrant student Tra:

Int: How good would you say you are at maths?
Tra: I don't know.
Int: Tell me what you think.
Tra: Like in the class I think I'm OK, I think that I don't believe I am not the best I am medium class, as in another class there are other people who are higher and there is students who is clever, so I would put myself in the middle.

So Tra takes a wider perspective on her level than just what happens in this class. Then more about levels and standards:

Int: Can you describe a situation in the maths class that was a bad time, a bad experience?
Tra: Not now but it was last year, I always do my work but there is no test and there isn't really test just assignments and they don't really show you what level you are. And I was so quiet as it was depressing as no one noticed you.....

She was used to doing tests frequently in her home country but now she felt she had no chance to find out how well she was doing. This surely must be one of the most important aspects of learning in a new environment. The interviewer encourages her to enlarge on this point:

Int: How good do you think your teacher thinks you are?
Tra: I don't know, no idea. It is hard to know what the teacher thinks.
Int: Tell me why it is hard to know.
Tra: Even though you might do well they might think something different about you.

Int: Can you tell me a bit more about that?
Tra: Sometimes it is kind of like people notice different people you are drawn to different people, so like sometimes even though you are good you are kind of like a shadow.

This metaphor is a powerful one particularly for someone working in her second or perhaps third language. She sees herself as a shadow, existing, but not really present; able to be ignored. How much does this contribute to teachers' under-rating of girls' achievements? Perhaps because of the language barrier she didn't say a great deal in class? Perhaps in her culture it was considered impolite particularly for girls to talk in class or to become obvious, or to stand out by being different? Perhaps she too wanted to merge with the class and in some way not be noticed?

Int: You feel like a shadow in the class?
Tra: I used to.
Int: But not any more?
Tra: No.
Int: Tell me why it has changed.
Tra: I used to be quiet and do my work and stuff but now I kind of like to communicate a bit and sometimes I join in a bit but not all the time as it is a bit of fun and a bit of work and make it good, better in the maths class, no work and play, it is a mixture and is good and I feel more comfortable with the talking.

Apparently she has made a transition and has adapted her behaviour to the new classroom culture. She feels more comfortable. But she's not communicating for the purpose of generating better understanding, or to clarify a half-formed idea, or to contribute a point of view in a class discussion. It is only a bit of fun, and play, to be contrasted with work. But it is also her way of finding her new identity in the case. She feels she has 'joined the crowd' perhaps. What makes this attitude even more poignant is that she was the top student in her class when we interviewed her. How long did that continue, one wonders?

The interview extract with Gor, a Year 7 student, reveals another aspect of the interaction with the teacher.

Int: Do you ask the teacher for help?
Gor: I didn't need his help for the work that we were doing.
Int: Did he used to come round and check anyway?
Gor: Yes, he gave us more advanced work when we were doing fractions, like the kids that were only normal, the kids in our class were just doing multiplying and addition and subtraction, and he gave me division, a few other kids got division.
Int: Did you enjoy that?
Gor: Yes.

Gor clearly appreciates being given work that he feels is more appropriate for his mathematical level, calling it 'more advanced work'. He also didn't think of himself as 'normal' in that class, perhaps over-rating his achievements? He didn't need to ask the teacher for help but he was clearly pleased that the teacher had recognised his achievements. But the next part of the interview reveals another aspect to his not seeking help from the teacher.

Int: What happens when you ask for help?
Gor: Sometimes the teacher would think that you don't know so much if you ask for help, but if its for like you don't understand the question, I have asked him for a few things.
Int: So you think that if you ask questions that is implying you don't know as much as you do?
Gor: Yeah.

Immigrant students, like all students, recognise the social and public nature of learning in a mathematics classroom. Perhaps though they are even more aware of the need to understand the subtle signals that trigger teacher perceptions and determine teacher behaviours towards them, and ultimately teacher assessments of them. The transition experience clearly heightens the awareness of the subtle interpersonal body language and other forms of communication.

6.2. *Teaching Approaches*

Most immigrant students have been to school before, although not all. If they have then clearly they have already formed conceptions of what school mathematics teaching and learning is about. In a new situation this is likely to be different. We have seen above that the interactions between teacher and students may well be different. What about their feelings about the teaching approaches used in their Australian classrooms?

Here is an extract from the interview with Ste, a Year 9 immigrant student. Firstly the interviewer asks about his friends:

Int: So is there anyone in particular that you like to work with in the class?
Ste: Yes, J (not an immigrant)
Int: Why do you like to work with him?
Ste: He is good at maths.
Int: How good would he be compared to you?
Ste: A bit better.
Int: You like to work with someone that is better than you?
Ste: Yes. Bit more knowledge.

but then the topic moves on:

Int: What is it like being in your class?
Ste: It's boring.
Int: Is it? Tell me a bit about that.
Ste: If we could build a maths lesson, like okay, we are just doing the questions, and the teacher writes on the board and we just do it, and when we have finished, and she explains it that's all. But if we could do a more like… (pause)
Int: Have you had this experience before, like have you done maths in a different way, have you an idea of how you would like it? (Pause)

Sounds like you would like a different structure for your class?
Ste: Yes, that is correct.
Int: Do you like the group work that you did?
Ste: Yes that was really good, I mean if the whole class works on one question, the question was asked, it is better, like when someone comes to the board, if we could do it all the time like this, someone comes to the board and explains the question it is a lot easier, if you don't know the question the whole class helps.

It is clear of course that not only was this an experience of a different teaching approach it was also one with which Ste could succeed. This is not to suggest that Ste was not succeeding at present, but only to point out that experiences of other teaching approaches carry with them expectations of success or failure. The transition to different teaching methods can be as difficult to handle as can the transition to different teachers. However from the perspective of researchers the transition to another teacher highlights some of their important characteristics for the students.

Reg, a Year 7 High Achiever, reflects this aspect in her interview, as they have just been discussing her previous teacher Mr T.:

Int: So you were happy with your teacher?
Reg: Yes.
Int: What about now, are you happy with your new teacher?
Reg: No. Don't tell her!
Int: So she is a bit different?
Reg: She is really different, like she doesn't explain as much but she gives you lots and lots of homework. Mr T explains it more and didn't give you as much.
Int: What was it like being in your class with Mr T?
Reg: Interesting. I would work more but have a bit of fun. But she goes 'do your work', 'you will have to stay in at lunch'.
Int: So she is a bit stricter?
Reg: Yeah, we were more free to do, like you learnt more, I think his methods were better.
Int: Can you describe a situation in maths that you liked with Mr T?
Reg: I like it in class when we do challenging problems in groups, but I always get them really quick and then I have to explain it all to my friends, and they don't understand because I talk really fast, and they don't understand. But Mr T understands me.

Mr T clearly has a fan in Reg, but she has a more challenging situation now with her new teacher. We however can gain some more understanding of Mr T's approach from an extract of the interview with Dei who is another Year 7 immigrant student but in Mr T's class:

Int: And how good would your teacher think you are?
Dei: I don't know. He teaches at different levels, like for the new girls who have just come from Russia and everything, they have a different level, they have different work to do. Like and the rest of the class he gives us a certain amount of work to do. Whoever finishes first he gives us more challenging problems.
Int: And do you ask the teacher for help?
Dei: Everyone used to ask him for help so you had to wait in line.
Int: Did he explain things well?
Dei: Yeah he did.

Int: Can you give an example of a good learning experience with MrT?
Dei: I don't know. Hard to remember. He gives us different things each lesson.
Int: So you like the variety?
Dei: Yeah.

We can infer from other parts of the interview data that Reg was also from Russia but that she was not one of the newcomers mentioned by Dei. However she has now moved on to the other teacher with the other girls. It was also clear that those girls' backgrounds in mathematics were ahead of the rest of the students in Mr T's class, which was why he had to 'teach at different levels'. This is clearly understood and appreciated by the students in his class.

This of course is one more aspect of the transition experience for the students. With immigrants coming from so many different countries, the differences in experiences of previous teaching, of the level of mathematics, of the work demands etc. all mean that the mathematics classroom is a situation characterised by difference. Perhaps it takes a certain kind of teacher with a variety of approaches to cater to all the needs and expectations. Mr T is obviously one example.

6.3. Peers and Their Influences

All students know that the relationships with one's peers can be crucial for survival in the classroom and in the school. The immigrant students in particular learn many things from their classroom peers. But they often have to deal with difficult situations and to learn from them.

Here is Tra again:

Int: What's it like being in your class?
Tra. Like everyone I feel doesn't accept me and I want to work and get ahead, and they are yelling at me.
Int: You think that yelling is in fun?
Tra: Yes.
Int: Who yells at you?
Tra: All the boys and girls as they don't want to work.
Int: And they say hey you are working too hard?
Tra: Yes.
Int: What do they say?
Tra: Nerd head, and they are friendly.
Int: But its not a put down?
Tra: No, but I just yell back and start to annoy them.

So she has learnt some strategies to deal with the situation. But deep down what does she want?

Int: Do you feel comfortable in the maths class?
Tra: I just want to blend in with the crowd.
Int: So you feel that you stand out a little bit from them more than you want to?
Tra: Yes. I like the attention and stuff but sometimes it's a bit much, you want to fit in with them.

But that is difficult in the public arena of the classroom. What about working more collaboratively with someone?

Int: Is there anyone you like to work with?
Tra: Someone who understands, a close friend, so when you do well they don't get jealous or anything they help each other and that kind of friend, and when I am in trouble she will stick by me and stuff.

Having clarified her criteria, but not answered the question, the interviewer tries again:

Int: Is there anyone you like to work with?
Tra: Someone challenging, he doesn't seem to try hard, Bha (another immigrant student). I ask him to try hard and do his best so that I can challenge him and I want to see what level I am compared to him.
Int: So you like to work with him?
Tra: I just want to compare marks.

So it is not just collaborative working that she seeks in order to get accepted more. Once again Tra seeks help with clarifying how well she is doing, and this time from the peers, the other source of information besides the teacher. Perhaps she values the feedback from the peers more? Perhaps also the information from the peers will also tell her more things about the cultural norms of her new school?

Gor (Year 7) talks about some of the same issues in these extracts from his interview:

Int: What's it like being in your class?
G: I don't know, they call me a square because I know more than normal kids. I don't like them calling me a square.
Int: Tell me a bit more about that. Who calls you a square?
G: All the kids who don't know much.
Int: Why do you think they do that?
G: I don't know, because they are jealous or...
Int: How do you cope with that? Do you say anything back?
G: I just ignore it.
Int: Does it get worse if you ignore it or...
G: They are just like, sometimes I say thanks and I like them saying it, so it sounds like I like them saying it, so they will probably stop saying it!

Gor has a different coping strategy from Tra's but it seems equally effective. Gor has a particular view of himself, as we saw in the previous extract above. Here he says 'I know more than normal kids' which comes as much from the feedback he gets from the peers as from the teacher. Once again the interviewer probes about his working with particular friends.

Int: What happens when you get stuck on something in maths, what do you do?
Gor: I keep going until I finish it.
Int: So you persist with it yourself?
Gor: Yes.

So does Gor have any concept of being 'stuck'? Is he perhaps a 'loner'?

Int: Do you check with anyone else?
Gor: If I am stuck I will check with some other kids.

Int: Who would you check with?
Gor: My friend who is as good as me in maths.
Int: Who is that?
Gor: Thi (another immigrant student).
Int: So he is equal to you?
Gor: Yes.
Int: Do you work with him a lot?
Gor: I do the same things, I don't always work with him though, but we ask each other questions a lot.

Choosing particular peers seems to be a sensible strategy to blend the needs for survival, a sense of belonging, and a sense of what norms and standards are important. In this next extract, the Year 9 student Mar is talking about the particular friends she likes to work with:

Int: Is there anyone in particular you like to work with?
Mar: Yes, I like to work with my friends Rula, Kylie, Lina.
Int: So you chose to work with them as they are your friends?
Mar: Yes they are my friends and they are pretty good at maths so, and since we get on well we do things faster like problem solving.
Int: Would you say they are better than you or just as good as you?
Mar: Oh no, not as good.
Int: You would be the best?
Mar: Not the best. Rula is not that good, but they are alright.
Int: How good do you think they think you are?
Mar: Oh, they think I'm good.
Int: Do you think that they think you are better than you are?
Mar: Yeah.
Int: Does that make sense?
Mar: To me it does.
Int: Why do you think that because that seems to be a common thing?
Mar: Because we sort of have these standards, because I would think that a good mark would be 90% and they think 70 is pretty good, and so when I get 80 they think I am really, really good, but I am aiming for higher.
So we have different standards.
Int: Where did your standard come from?
Mar: I don't know. Probably out of Russia, and just like a perfectionist, not completely, but I try to be as good as I can get. When I feel that I can do better then I am pretty upset. Like I haven't put enough work into it, disappointing.
Int: Some students I have talked to have felt pressure from other classmates when they do well, they get called a square.
Mar: I don't worry about that, my life is, sort of, like I don't live my life for those people, they are obviously not my friends if they say that. And if they are just joking I don't mind.
Int: So do you feel that pressure at all?
Mar: No, I probably would if I paid any attention to it, not really, my friends don't, maybe as a joke sometimes, I don't mind that.
Int: So you decided just to ignore it?
Mar: Yes. It's the marks at the end that is more important.

Standards again! Her peers are clearly her friends here, but perhaps friends in a mathematics classroom are not necessarily friends at home. They obviously serve various functions in the class, and she, like the other interviewees needs them to function at the level she has chosen.

The issue of standards keeps reappearing in the interviews, in different places and in response to different questions as we have seen. It has emerged in relation to the teacher's interaction patterns, to the teaching approach, and to the choice of friendship peers. Several students spoke of the difference in standard of mathematical work at schools between their former country and Australia. Two example interview extracts follow, the first from the interview with a Year 7 high-achieving male student from the Ukraine:

Int. I keep getting a sense that in the Ukraine you worked at a different standard of maths, is that the case?
M. Yeah well I only went up to 3rd grade, so I can't remember.
It was harder the stuff I did then, when I came here the stuff
I did in grade 5 I did in grade 3 there.

The second is with a Year 9 low-achieving male student from Vietnam:

Int. What was it like doing maths in your own country?
Ty. In my country it is more difficult, when in year 7 we had equations and parallelograms or parabolas and bar graphs.
Int. So you did all that in year 7?
Ty. Yes
Int. And now you are doing it in year 9?
Ty. Yes.

but despite this the teacher had classified him as a Low Achiever.

So there are distinct differences in the levels of the mathematics curriculum which play out their effects through the students. This is well demonstrated by Dan, a Year 9 student from Georgia, who raised the topic himself following a discussion about who he liked to work with in the class.

Int: So there is no-one in particular that you like working with?
Dan: In maths I would rather not work with someone, rather work by myself.
Int: And what is the main reason for that?
Dan: Well I learnt the basics in Georgia right, and not like here, they wouldn't let you talk in class, you would be sitting down by yourself and working, working, as we have a lot more work in Georgia than this, a lot more, I mean we get maybe one exercise here, there we get ten times as much. We would have to sit down the whole class. I can remember in Grade 3 I was coming home and I had more work than I have for the whole of one day here, I would be sitting there until the evening to do that. If I had only enough work till it was not even dark, still light, I would be happy, because I would have ten minutes to go out or something. The school system there is very strict as well, and that is why I like to work by myself.
Int: So you have the experience of two different systems what do you think you prefer?
Dan: Well I think that the Georgian system is better in terms that you would learn more.

So it is not just a matter of the formal curriculum. The standard of work is perceived by the immigrant students through the class activities and the homework also. The teaching approach, the curriculum, the strictness or otherwise of the teacher, the behaviour of the peers, all contribute and become blended into a picture of the 'mathematical work-load' i.e. how much you are expected to work in class. This is clearly a crucial part of the mathematical practice in classrooms.

This is also an important idea in the transition stages of immigration. Whether its effects would continue is hard to determine but moving into a new classroom in a new country and finding that the mathematics being taught is easier than it was 'at home' must in one sense be a reassuring situation. It would be particularly so in comparison to the other subjects containing not only new content and new perspectives but probably also involving many more language complexities.

Another point worth making is that this experience would not have been shared by all immigrant students. It is certainly known that in parts of Europe, particularly the former Eastern Europe, and many Asian countries, the mathematics curriculum is more demanding and advanced than that in Australia, but there are many other countries for which this is not the case. Of course if the mathematics in the new situation is 'harder' than what one was used to at home, this could create a very painful and destructive learning experience.

6.4. Parents: Sources of Pressure and Support

The final source of influence and pressure to be considered is the parents, and this influence was clearly shown in many of the student interviews. Not only were there important differences between what they thought their mothers' and fathers' ratings of their achievements would be, but it was clear that they felt that their parents had generally high expectations for their performance in mathematics.

Here is Gor again:

Int: Tell me about why maths is one of your favourite subjects.
Gor: Because I am good at it. And in Grade 2 we had a good maths teacher.
She wasn't a maths teacher, (she was) a grade teacher and she taught us good maths and so did my dad.
Int: Your dad taught you good maths as well?
Gor: Yeah.
Int: What did your father teach you?
Gor: He taught me decimals, algebra, problem-solving.
Int: So you have had a lot of positive influences from your family?
Gor: Yeah. And recently my dad had some sheets, like on sets and unions.
Int: Is your dad a maths teacher?
Gor: No he works in a school, he's the cleaner there, but he still knows maths.
He came from Macedonia to Australia and he didn't have English so he had to work there.
Int: What did he do in Macedonia?
Gor: He was a building technician for bridges, he had to test them.

(Later)

Int: Your father would like you to be better or top?

Gor: Yeah, because in Macedonia when they were at school they would learn much more than what we learn here...

Int: So your dad is a bit concerned that the standard is a bit lower than in Macedonia?

Gor: Yeah. He says if I go there, in language, I would be very bad because I can just write, I know the alphabet, but not off by heart, I just know the letters, and if I go there I wouldn't be the top of the class, because they learn more maths than here....

Int: Is there anything else you would like to tell me, like what would be important for teachers to know?

Gor: I think they should make the maths like three groups, three classes, one where all the students who don't know nothing should go, another one where they already know it, they just need revision, so they just get the questions, and another where you learn newer things for a higher standard.

Int: And you would like to be in that latter one so that you are more challenged?

Gor: Yeah, because all we are doing in Year 7 is just revising.

As we noted before Gor has a particularly confident view of himself and this comes into play with his 'solution' for the teachers. Before dismissing it as a rather immature and biased idea though, perhaps it is worth recalling the skills of Mr T about which we have already commented. One wonders whether Gor would have appreciated his class as opposed to the one in which he now finds himself.

Jam, a Year 9 student also indicates the combination of pressure and help that the parents can provide:

Int: Do you think in Russian when you are doing maths?

Jam: Sometimes, because my father is an engineer and he helps me a lot.

Int: And he would speak to you in Russian when he helps you?

Jam: Yes....

Int: You said you were a 4 in maths. You said that your father would only rate you as a 3, so your father doesn't think you are as good as you are?

Jam: Yes, because when we came here, like our level went down, like we didn't do anything in primary school and so on. And when father compares it to Russia he would put me as a 3.

But what of the parents themselves? We were not able to arrange many interviews with parents but some were possible and were revealing. There was the expected support, and level of expectations. Here is an example from part of the interview with the mother of an immigrant student:

Int: How well would you like him to be doing at maths (Out of 5)

M: Yeah I would like him to be a 5. Like every parent.

Int: And do you tell him that?

M: Yes I do, I expect – I told him, I expect you to be the best.

Int: The best in the class?

M: Mm sort of, one of the best, not the best in the class, but one of the best.

Int: And how has he been going in maths, since he came over, from overseas?

M: Oh, because in Russia, with maths, a little bit higher than here, I think, in all Europe. A little bit higher, so he just was in, and was a little bit problem, because he didn't understand language well, so I – he didn't understand questions...but the maths he could understand....

Int: So when you came here, the maths is a bit easier. How many years difference do you think it is?

M: I think 2 or 3 years.
Int: That much do you think. Do you think that means the maths is easier here?
M: Easier, much easier, because my little son, he is now in grade 6, and back in Russia, I remember what he was doing – some things in grade 4, grade 3.

This then confirms or at least supports some of the students' perceptions of the difference in levels and standards. But more is to be revealed:

Int: Some people say in response to that, yes they might learn it there but it's more rote learnt, and maybe they don't understand as much. So I'm interested to see what parents think.
M: Mm and another thing, here they are more logical than there.
Int: More logical?
M: More logical they ask questions, like what do you think about this and that....not to have just 5 plus 5, you know....just more logical, I think so. Yeah.
Int: So which system do you think is the better?
M: Here.
Int: Because of that?
M: Mm.

Presumably it was the parents' choice to migrate to Australia, not the children's, so it might be expected that a parent might have a different view about the new education system than their children. However given the similarities we generally found between the students' views in our interviews and their perceptions of the parents' views we must say that this parent's view was rather unusual.

7. DILEMMAS AND DECISIONS

How then should one interpret the various experiences and challenges which these immigrant students have so eloquently revealed to us? Note of course that there are different responses from the students representing different backgrounds, and also perhaps different foregrounds. As was said above the immigrant classroom is characterised by differences, and perhaps the most appropriate way to reflect on these differences is to consider them through the necessary dilemmas that students face, and the decisions that they have to make, as they go through this transitional mathematical practice. Separating these out for discussion purposes is always artificial because in the real situation they are always interacting, and are merely different perspectives on the same situation.

Firstly concerning the teacher, the defining characteristic of the teacher's role is that of a guide, or even as a leader, through the transition situation in class. One clearly looks initially to the teacher as the source of support, help and advice. It is their job, their task, their role to be that leader and guide. The dilemma or decision for the immigrant student however is whether to follow that lead, or to remain in some sense independent.

Where does this independent alternative come from? As has been seen, these secondary students have already experienced school elsewhere and therefore they already have images and pre-conceptions of how the teacher should behave. They may not always have enjoyed their previous learning experiences but they were generally part of their enculturation and induction experiences into their first culture. As such they would have accepted them, and learned to survive in them, or even succeed in them. To remain independent in the new situation is to rely on one's previous knowledge of how to behave with teachers in order to survive and do well.

What might be lost if one 'follows' the new teacher? There is certainly the loss of the sense of standard, which kept on emerging in the interviews, so one has no sense of how one is getting on, whether one is doing enough work, etc. More importantly, the teachers in their gate-keeper role, make many assessments which determine one's passage through the school structure. They are the conduits to success in the final examinations and to whatever career one wants. In some way the assessments made by the teachers model the later and final examinations, and even though there may be few opportunities to discuss openly the criteria, both overt and covert, by which the assessments are made, the unwritten didactical contract between teacher and learner says in effect to the student 'trust your teachers, they know best'. The key dilemma for the immigrant student is therefore whether they can trust their new teacher's judgements about them. As Tra put it earlier 'Even if you do well they might think something different about you!'

Secondly, concerning the mathematics, the defining characteristic emerging from these interviews is 'mathematics as a work practice'. The level and standard of this work was a prominent feature in the interviews, and relates to the previous point in that the immigrant student is always referring back to their previous experiences, which faces them with a dilemma. The dilemma is whether to set one's own challenges, based on one's previous experiences, or to accept the new levels and standards. This sample of immigrant students was perhaps rather biased, in that many of them referred to the level of mathematics as being lower in Australia than in their home countries. This meant that there was an opportunity for them to relax and not work very hard, with the result that some like Tra and Gor wanted to be challenged, and had set their own challenges as a result of not accepting the new lower levels and standards. They were responding to the opportunity to get a head start in the mathematical 'competition'.

Again, what might be lost if one chose to follow the new ideas on standards, if one chose to believe in and adopt the school and class norms? Of course one thing that would be lost would be the opportunities for the head start. But in the case of these students, another might be a loss of the respect of one's family. One can imagine the battles with the parents, who are also aware of the different standards but who have high aspirations for their children. To what extent might an immigrant student go against their parents' wishes, particularly in a new country? That is quite a dilemma.

Thirdly there are the peers. Here the defining characteristic is the peers as models, even as role models. The dilemma then for the new student is does one become assimilated by, or does one challenge, the models? Many times in our interviews the students referred to the ways the peers treated them, particularly if they were perceived to be working too hard. Two images frequently seemed appropriate when considering that evidence, firstly that of the 'union officials', i.e. the classmates who have negotiated the terms of the didactical contract for working in that classroom with that teacher, who they see as the 'boss'. They would not be happy with a new student 'rocking the boat' and giving the impression that more work could be done.

The second image is that of the apprentice/master relationship. The learner/teacher relationship is often described this way in the literature, but here it has relevance for describing the 'newcomer / old hand' inter-student relationship. It may be convenient to think of the teacher as the leader and model for the immigrant students, but the reality is that the students want to be like the other students. So it is the 'old hands' amongst the classmates who are educating the newcomer about the rules of engagement, the level of work to be done, the extent of collaboration etc. The old hands are the 'masters' and the newcomers are the 'apprentices', albeit already educated ones. The main difference from the original idea of the metaphor is that generally there the apprentices wanted to be apprentices. Here the decision is still to be made by the immigrant students: do they want to become apprentices to the old hands among the students? Certainly the 'masters' in the classroom know that they are the masters, and even can demonstrate this by violent means such as bullying or teasing if necessary, although we had no direct evidence of these.

Of all the influences this one seems to create the most dilemmas. Does the immigrant student try to fit the peers' norms or does one stay with one's preferences and instincts? Does one bow to the peers' pressures, or does one take the wider view and, perhaps even slightly good-humouredly, dare to be different? In this case it is easy to see what would be lost if one chose not to follow the norms. Basically one would forever remain an outsider, a difficult situation for any adolescent, even a good humoured one. But what is lost by following the other path, and by trying to become assimilated as much as possible? This means giving up much of what had helped one succeed in the previous situation, particularly in the nature of the type and extent of the mathematical work done. It means giving up one's previously learned meanings for classroom actions and behaviours. What was appropriate before is now no longer acceptable. What was not acceptable behaviour previously is perhaps now appropriate. It is risky.

Dan gave us an excellent example of his dilemma when asked about whom he worked with in class:

Int: Is there anyone that you work with now and again?
Dan: What do you mean work with? Do I get my friends to help me?

The mathematical work practice that Dan was used to formerly did not allow for collaborative activity.

Such are the dilemmas and decisions facing the immigrant learners in the new mathematics classroom. Perhaps the more conservative students will stay with what they know best, trusting their knowledge of favoured teachers and approaches 'back home', keeping in their heads the 'real' mathematical standards that are ingrained in their memories, and boldly holding out against the strong peer pressures to conform. Perhaps the more anxious ones will decide there is more to be gained by joining rather than beating the peers, and will quickly learn from the 'masters' how to negotiate with each teacher and in each situation. Perhaps the braver and more adventurous ones will be prepared to take more risks with both teachers and peers and even seek to make their mathematical norms, standards and practices more acceptable to the 'locals'. Each immigrant student has potentially the power to change, however slightly, the accepted rules of each situation, rather than just become assimilated into the host culture. Whether they will do that depends primarily on them of course, and on the strength of their background and support. It also depends on how culturally responsive the teacher, the curriculum and the peers are and how prepared they are to accept difference in mathematical practices. The transition experience is not only the student's, but also belongs to the other co-constructors of the classroom culture.

REFERENCES

Atweh, B., Cooper, T. and Kanes, C. (1992). The social and cultural context of mathematics education. In B. Atweh and J. Watson (Eds) *Research in mathematics education in Australasia 1988–1991* (pp. 43–66). Queensland University of Technology: MERGA.

Australian Educational Council (1991*). A National Statement on Mathematics for Australian Schools.* Carlton, Australia: Curriculum Corporation.

Bishop, A.J. (1994). Cultural conflicts in mathematics education: developing a research agenda. *For the Learning of Mathematics. 14* (2), 15–18.

Bishop, A.J. (1995). Western mathematics, the secret weapon of cultural imperialism. In B. Ashcroft, G. Griffiths & H. Tiffin (Eds) *The post-colonial studies reader*. (pp. 71–76). London: Routledge

Bishop, A., Brew, C., Leder, G. and Pearn, C. (1996). The influences of significant others on student attitudes to mathematics learning. In L. Puig and A. Gutierrez (Eds) *Proceedings of the Twentieth Conference of the International Group for the Psychology of Mathematics Education.* Vol 2 (pp. 89–96) Valencia, Spain: University of Valencia.

Bishop, A.J., Leder, G., Brew, C. and Pearn, C. (1997). Researching cultural issues with NESB secondary mathematics students: 'In my country its more difficult'. In F. Biddulph and K.Carr (Eds) *People in mathematics education* (pp. 96–102). Waikato University: Mathematics Education Research Group of Australasia Inc.

Dawe, L.C.S. (1983) Bilingualism and mathematical reasoning in English as a second language. *Educational Studies in Mathematics, 14*, 325–353.

Ellerton, N. and Clarkson, P. (1995). Language factors in mathematics education. In B. Atweh and J. Watson (Eds*) Research in mathematics education in Australasia 1988–1991* (pp. 153–178). Queensland University of Technology: MERGA.

Fennema, E. & Sherman, J. (1976). Fennema-Sherman mathematics attitude scales. *JSAS: Catalogue of selected documents in psychology, 6* (1), 31 (Ms No 1225).

Fennema, E., Wolleat, P. & Pedro, J.D. (1979). Mathematics Attribution Scale. *JSAS: Catalogue of selected documents in Psychology, 9* (5), 26 (Ms No 1837).

Fraser, B.J. (1990). *Individualised classroom environment questionnaire: Handbook.* Hawthorn, Australia: Australian Council for Educational Research.

Howard, P. (1996). Aboriginal educators' views concerning the learning and teaching of mathematics. In P.C. Clarkson (Ed) *Technology in mathematics education: Proceedings of Nineteenth Annual Conference of the Mathematics Education Research Group of Australasia* (pp. 298–305). University of Melbourne: MERGA.

Leder, G., Bishop, A.J., Brew, C. & Pearn, C. (1995a). Learning mathematics in context. In J. Wakefield & L.Velardi (Eds) *Celebrating mathematics learning*. (pp. 438–443). Brunswick: Mathematical Association of Victoria.

Leder, G., Bishop, A.J., Pearn, C. & Brew, C. (1995b). The mathematics of success: beyond questionnaires. In *Proceedings of Annual Conference of Australian Association for Research in Education.* *blhttp://www.aare.edu.au/index.htm*bg

Leder, G., Rowley, G., and Brew, C. (1995c). Second language learners: help or hindrance for mathematics achievement? In R.P. Hunting, G.E. FitzSimons, P.C. Clarkson and A.J. Bishop (Eds) *Regional collaboration in mathematics education* (pp. 425–434). Melbourne: Monash University.

Leung F.K.S. (1998) The traditional Chinese views on mathematics and education: implications for mathematics education in the new millennium. In H.S. Park, Y.H. Choe, H. Shin & S.H. Kim (Eds) *Proceedings of the ICMI-East Asia Regional Conference on Mathematical Education* (Vol. 1, pp. 69–76). Seoul, Korea: Korea Society of Mathematical Education.

Mehan, H. (1996). Beneath the skin and between the ears: a case study in the politics of representation. In S. Chaiklin & J. Lave (Eds) *Understanding practice: perspectives on activity and context* (pp. 241–268). Cambridge: Cambridge University Press

McDermott, R.P. (1996). The acquisition of a child by a learning disability. In S. Chaiklin & J. Lave (Eds) *Understanding practice: perspectives on activity and context* (pp. 269–305). Cambridge: Cambridge University Press.

Moscovici, S. (1976). *Social influence and social change*. New York: Academic Press.

Parsons, J.E. (1981). Attribution, learned helplessness and sex differences in achievement. In S.R. Yussen (Ed.) *The development of achievement*. New York: Academic Press.

Pearn, C., Brew, C., Leder, G., and Bishop, A.J. (1996). Attitudes towards mathematics: what about NESB students? In P. C. Clarkson (Ed.) *Technology in mathematics education: Proceedings of Nineteenth Annual Conference of the Mathematics Education Research Group of Australasia* (pp. 445–452). University of Melbourne: MERGA.

Thomas, J. (1995). Bilingual students and tertiary mathematics. In B. Atweh and S. Flavell (Eds) *Proceedings of Eighteenth Annual Conference of the Mathematics Education Research Group of Australasia* (pp. 511–516). Northern Territory University: MERGA.

CHAPTER 4

THINKING ABOUT MATHEMATICAL LEARNING WITH CABO VERDE *ARDINAS*

MADALENA SANTOS AND JOÃO FILIPE MATOS
Centro de Investigação em Educação
Faculdade de Ciências da Universidade de Lisboa

The research reported in this chapter takes part of the results of the project *Cultura, Matemática e Cognição – Pensar a Aprendizagem em Portugal e Cabo Verde*[1], especially in what concerns understanding the idea of 'learning as an integral part of generative social practice in the lived-in world' (Lave & Wenger, 1991, p.35) together with the perspective that addresses 'learning as increasing participation in communities of practice' (p. 49). The analysis of these ideas led us to try to understand the meaning of participation in a social practice (and therefore in a community of practice). Our goal was to look into the ways (mathematics) learning relates to forms of participation in social practice in an environment where mathematics is present but that escapes the characteristics of the school environment. Because we believe that culture is an unavoidable fact that shapes our way of seeing and analysing things, we decided to look at a culturally distinct practice and that constituted a really strange domain for us: the practice of the *ardinas*[2] at Cabo Verde islands in Africa.

In order to address the research problem we looked for analytical tools that we believe are coherent with the theoretical perspective drawing from Lave and Wenger (1991) and our need to understand the idea of social practice: (i) the explicit presence of rules in the *ardinas* discourse led us to Wittgenstein (1992/1953) and Goffman (1991/1974); (ii) the need to clarify the relationship between the ways that *ardinas* use mathematical objects within the practice of selling newspapers and the school mathematics they are supposed to know led us to address the connections between competence and 'taxonomic features of knowledge' (Julien, 1997) and to the idea of 'thinking as internal conversation' (Restivo, 1998).

[1] Culture, Mathematics and Cognition – Reflecting about Learning in Portugal and Cabo Verde. This project was supported by Fundação Ciência e Tecnologia under contract PRAXIS/PCSH/C/CED/146/96.

[2] *Ardina* is the Portuguese name given to those people who sell newspapers in the street. This was the way newspapers were sold for example in Lisbon until the eighties.

G. de Abreu, A.J. Bishop, and N.C. Presmeg (eds.), Transitions Between Contexts of Mathematical Practices, 81–122.

In this chapter we first set the scene presenting a full description of the practice observed in Cabo Verde; the subsequent analysis takes advantage of short episodes depicted from the observations made; finally we conclude with a summary of our main findings.

1. A GLANCE ON THE *ARDINAS* PRACTICE

The practice of the *ardinas* has naturally a history that relates to the evolution of the country and to the society where it develops. The data collection that supports this research was carried out in two time periods: the first phase was from March to June 1998 and the second phase during the month of March 1999. This way it was possible to enlighten the historical transformation of the practice, mainly through focusing our attention on the transitions experienced by the participants. Focusing our analysis on those experiences helped us to understand better the forms of participation of the *ardinas* in that practice, its role in the learning that came out of that participation as well as in their use of mathematics. This research process brought also to the front some other issues that pushed us to reflect upon the research process itself within the field of mathematics education research.

Therefore, it seems important to give a picture of the life that we could share with the *ardinas* during the time of data collection, trying to make explicit the relationships among people, between people and activity and the lived-in world. We begin this part with a brief and global presentation of the *ardinas* involved in the study. In a second step we will present a description of the practice observed taking into account the time sequence of its development.

1.1. Who are the Ardinas?

The *ardinas* are young boys aged between 12 and 17 years that sell newspapers in the streets of Praia (the capital of the Republic of Cabo Verde). In 1998 there was just one national newspaper (called *O Tempo*) but from January 1999 there appeared a new one (called *O Espaço*). Both of these newspapers come out once a week and are written in Portuguese[3]. The group of *ardinas* who used to sell these two newspapers was variable (19 in 1998 and 32 in 1999). Only 9 *ardinas* from the 1998 group were carrying on this practice in 1999, and there was no formal link to the institutions that owned the newspapers.

One of the newspapers (*O Tempo*) was trying to implement a selling system based on the shops such as coffee shops or stationary shops but with very low success. In fact the population did not adapt to this way of buying newspapers, so selling news-

[3] Portuguese is the official language in Cabo Verde. Therefore, it is for example used in the school and it is the language of the newspapers. However, Creole is the spoken language used in everyday activities. Far less people are really fluent in Portuguese.

papers in the city of Praia was totally dependent upon the availability and interest of the young boys to volunteer for selling. Some of these boys also were in charge, both in the past and in the present, of selling Portuguese newspapers, and especially a particular one on sports very popular in Cabo Verde. This one was on sale on the street but on a different day. There were also two other *ardinas* who did not sell national newspapers because they were dedicated to selling two Portuguese weekly newspapers of general interest whose contents are more in the field of politics. These two young boys had already a small number (but a rather constant number) of clients and the places for selling were clearly different from the others. In most cases these *ardinas* were taking the newspapers to the offices and hotels in the city. The relationships between these two *ardinas* and the others were not very strong and frequently they did not even say hello when they met in the street.

There was no external sign (such as a special t-shirt, a bag or a cap) that could be one to identify the *ardina* except the fact that he was carrying a number of newspapers under his arm. However they were careful in the way they dressed on the days of selling. They managed to be clean and it seems that they tried to maintain a certain combination of clothes especially for that activity. Some of them had special care with clothing in order to have access to certain places of selling (for example, in official departments).

Some of the boys started selling newspapers just prior to the data collection for this project (March 1998) but others had already been selling newspapers in the street for about six years. Most of them were *ardinas* because they wanted to get some money to help the family (*'to help my mother'* as they usually said).

Because data collection was carried on in two phases we could identify differences in these two periods. In 1998, the group of *ardinas* was living in two places, 9 *ardinas* were living at the Eugenio Lima area in Praia, which was one of the most problematic places in the city mostly with inhabitants coming from the rural zones looking for a job in Praia. There was another group of 10 *ardinas* coming from S. Martinho, a small village close to Praia. In 1999 the group was enlarged with boys coming from Praia, and those from S. Martinho started leaving this activity[4]. The growing of the number of *ardinas* in the group was slow; for example, during the month of March 1999 we observed the integration of only one new *ardina*.

1.2. Ardinas' Practice

In the two periods of data collection (1998 and 1999) some common aspects were observed; but we could also identify different aspects. The work of the *ardinas* was divided into three different phases: (i) receiving the newspapers, (ii) selling, and (iii) paying back the money to the newspaper agency. The organisation of these three

---

4 Several factors contributed to the fact that these boys abandoned the practice of selling newspapers in the street: some of them had the opportunity of working in S. Martinho helping the construction of infra-structures in the village organized by the local authorities.

phases was necessarily connected to the instructions of the directors of the newspapers but the *ardinas* positioned themselves in that organisation in their own way according to several facts (that then become more visible). This means that in this practice we could identify a pre-existent structure or arena, in Lave's (1988) terms, which was redefined in the local context of action and that develops with the participation of the *ardinas* (and that certainly would be different for other *ardinas*).

1.2.1. In 1998

Every Friday morning, in the main building of *O Tempo* agency, the newspapers were delivered to Disidori, the man who was responsible for the whole process of selling[5], returning the non sold newspapers and payment. In order to have the newspapers sold Disidori distributed them among the *ardinas*; this operation took place at the door of the agency. The number of newspapers distributed to each *ardina* was negotiated and in most cases the number of newspapers varied between 50 and 150 for each. During the **distribution** Disidori wrote down in a list the names of the *ardinas* and the number of newspapers distributed to each one. This list was the reference document for the final phase when the *ardinas* were paying back to Disidori after selling.

The participation of the *ardinas* in the activity of selling was based on their will to do that. The link of the *ardinas* to the newspaper agency was very informal assuming a very personal character in relation to Disidori (more than to the agency); there was no penalty and no need for justification if the *ardina* decided not to show up for selling. If he decided afterwards to come back for selling he knew that he could do it (notwithstanding that he could not have immediately available the number of newspapers he wanted to have). On his side, Disidori had a link to the administration of the newspaper, which was made visible to all when he signed a document against the delivery of the newspapers (which made him responsible for the payment to the administration). Besides that, Disidori received a fixed amount of money (that he recognised as the payment for assuming the responsibility) plus a part of the money of selling each newspaper. We can say that there was a mutual dependence among the *ardinas* (in order to get money they had to sell the newspapers), Disidori (to get more money he had to be sure that the *ardinas* really sold) and the administration of the newspaper (to sell the newspapers they needed the help of Disidori and the *ardinas*).

The group of *ardinas* changed over time. After a varied period of time some of the *ardinas* abandoned the activity of selling newspapers. Usually these boys got involved in other activities (for example, serving the army, getting a job or emigrat-

---

[5] This adult had been one of the *ardinas* when he was a boy and for some years he was involved in the work with the *ardinas* but in no way that had not much visibility in institutional terms. He had not a working place in the agency of the newspaper and he only worked for the newspaper on the day when it was being sold. Besides that he lived in S. Martinho and had friendship links and even family links to some of the *ardinas*.

ing to another country). New boys then came to substitute for those who dropped out and generally these newcomers were accompanying a friend or someone from his family who was already an *ardina*. In the case of these new *ardinas* Disidori had the last word for their acceptance in the group. Disidori tried to know the *ardina* and talk to his family. He also used to choose one of the old-timers to take responsibility for the newcomer – 'to teach him and to protect him' as they said. In the distribution of the newspapers among the *ardinas* what happened was that the old-timer received all the newspapers (for him and for the newcomer) and it was his job to give a small number of newspapers to the newcomer (first 5, then 10, 15, etc.) as he was selling.

Immediately after receiving the newspapers the *ardinas* ran very quickly to the usual places for **selling** in the city; their goal was to try to sell all the newspapers during the day. Some of the *ardinas* tried to maintain their own place of selling. However, those places varied during the day according to the rhythm of selling and the rhythm of the city (namely, at the street in the rush time, at the working places on the working schedule, close to the restaurants at lunchtime). The price of the newspaper for the customer was 100 escudos; by the end of selling, the *ardinas* should pay to Disidori 87.5 escudos per newspaper sold and give back the non-sold newspapers. These amounts were defined by the newspaper administration. During the day most of the *ardinas* spent some time at the Square of the city or in nearby streets given that these were the places where selling was more common. In fact this is the area where most of offices, banks and public services, coffee shops and markets are located. On the other side this is the zone of the city where some local people develop their activity selling sweets and pottering on the street. Because those people stayed on the Square during the day, the *ardinas* got a close relationship with them which is in fact useful to both parts: the *ardinas* ask those people to keep a number of newspapers for them avoiding to have to carry a big number of newspapers, they exchange small coins in order to facilitate the change to customers and together contribute to attract potential customers.

Besides the strategic role of that interaction in the integration of the newspaper selling into the socio-economic life of the city, the Square was the place where Disidori stayed for long periods during the day of selling. He also walked around to the different places where the *ardinas* were selling in order to check how the process way going. Some time after the distribution of the newspapers by the *ardinas* Disidori went to the Square carrying with him a set of newspapers for the possibility of those *ardinas* who were in the school (and because of that could not come to the distribution of the newspapers at the agency) or that he could distribute to those who sell very quickly and ask for more newspapers. The Square was the main point of convergence of the boys at several moments during the day: (i) at lunch time, those who did not approach the restaurants to sell, stay and rest for a while, (ii) when they finished selling and came to pay back to Disidori.

Those *ardinas* who were in a beginning phase of learning the practice of selling usually kept close to an old-timer who was responsible for them. On one hand, this was because the newcomers received a small number of newspapers (which they had' to pay back to the old-timer before receiving more newspapers for selling). On the other hand, it was within the observation and interaction with the old-timer that the newcomer started to understand important aspects of the activity (how to handle money and newspapers, which were the good places for selling, how to address the customers). Besides that, the newcomers learned also to identify potential situations of risk. Because the number of newspapers that the newcomer received was increasing, he could start going farther and farther away from the old-timer and this also contributed to his self-confidence and gain of courage to go through different streets, getting independent of the old-timer.

As the *ardinas* finished selling the newspapers (or got tired of selling) they started showing up in the Square where Disidori was awaiting them for the **payment**. The *ardinas* approached Disidori, they said how many newspapers were left or how many they had sold, Disidori made the calculation (87.5 times the number of newspapers sold) with his hand calculator, he showed the result on the screen to the *ardina* who then gave him the money. Sometimes those *ardinas* with more independence made their own calculation (with their calculator or Disidori's calculator); Disidori trusted their calculation but generally he confirmed by himself the number of newspapers left and returned by the *ardina*. This moment was lived in groups by the *ardinas*; the newcomers also participated in the process (even if they received newspapers from an old-timer and pay him back directly). Several operations were in progress: some of the *ardinas* were counting the number of newspapers left for returning, others were counting and organising the money according to the value of bills and coins, they delivered newspapers to Disidori, observed the calculation, give the money to Disidori. The environment could seem confusing at a first glance; there was a lot of money in sight changing from hand to hand. However, observing in detail one could understand that everything was running in a certain order and this allowed that each *ardina* could see what was going on with the calculations (their own or those of a colleague). This was one of the opportunities for talking about the happenings of the day: for example, the customers who did not pay for the newspaper, the customers who returned the newspaper after reading it (making it possible that the newspaper was sold again), the *ardinas* who had lost money or made mistakes with the change or even the case of some *ardina* who did not show up to pay for the newspapers sold. So these were real moments of learning for all of them, both because of the stories heard and the behaviours observed (in their peers and in Disidori) face to face situations that were explained and discussed, and also because of the different activities that were taking place at the same time and the way the *ardinas* interpreted and solved their own problems. It was curious that after paying Disidori and keeping their own money, it was common that the *ardinas* did not go away immediately; they stayed observing what was going on or helping the others after seeing their case resolved. Another aspect that called our attention was the total absence within thc ordinary *ardinas'* discourse of any attempt to make explicit (verbal explanation or deliberately showing) their calculation strategies, checking processes or anything we could classify as some sort of mathematical conversation. We will look more closely to this issue in the section 2.1.

On the Square Disidori did not limit his activity to controlling the *ardinas* or receiving the money for the newspapers sold. He tried to give them suggestions of good places for selling at a certain time of the day, he made observations about the behaviour of potential customers specially those travelling in cars and stopping at the lights. Disidori was helping the *ardinas* to read the signs of the city and of its inhabitants in order to allow the *ardinas* to sell more quickly and more efficiently. But Disidori also talked a lot with the *ardinas* on subjects that had nothing to do with the process of selling. In fact he behaved as an adult who was feeling a responsibility for the education of those young boys taking care that they did not get involved in illegal street games, or making comments about the way they spent the money they earned as *ardinas*. This is in fact the natural and traditional way that in Cabo Verde adults feel and act in relation to children.

1.2.2. In 1999

During the period of the second data collection (March 1999) there were two different newspapers on sale in the city, both coming out once a week: *O Tempo* and *O Espaço*. Now, the majority of the 32 *ardinas* were involved in the selling of both the newspapers. However a small number of *ardinas* were linked only to one of the newspapers; this was due to (i) the fact that the new newspaper had a different day and place for **distribution**, (ii) the distribution was made by different people in the two newspapers, and (iii) there was an explicit instruction on the part of the administration of the newspaper *O Tempo* calling the attention of the *ardinas* to the fact that they should not sell both newspapers together. This instruction was given to the *ardinas* in a meeting that the administration of the newspaper promoted with them. This meeting was on 24 December 1998 and the *ardinas* were invited to a Christmas party. This was the moment when Disidori moved to a new position at the new newspaper *O Espaço*. This was also the opportunity for the local television making a piece about the *ardinas'* life. But the party was also the moment for the administration to announce the new rules: now for each newspaper sold the *ardinas* would receive 20 escudos (instead of the previous 12.5 escudos) but they had to go to the newspaper agency to receive the newspapers and go back there after selling in order to pay for the newspapers sold.

During this second period of data collection it was possible to identify different forms of participation. For example, in the case of Bétu, he had already been considered a good *ardina* in 1998.[6] Now, besides selling he was also being paid by the agency (only in *O Tempo*) for his work of putting together the pages of the newspaper after the printing process. And he received a large number of newspapers for selling (150 to 200) that he distributed to a small number of *ardinas* that he advised in the selling process (his brother, his cousin and a neighbour). In this new form of participation, the *ardinas* paid Bétu and he was responsible for the payment of the newspapers to the agency. But the other *ardinas* still saw Bétu as before: although he

---

[6] The *ardinas* use the expression 'to be a good *ardina*' to classify those who 'follow the rules' or who are honest, serious, relating it mostly with the payment process. We will show later on (point 2.2.2.) how this value is present in their practice and how it interferes with their interpretation of the profit.

had a particular relationship to the newspaper agency, they didn't recognise any authority over them.

Another case was Manu, one of the oldest *ardinas*, recognised by all of them as a good *ardina* and someone whom Disidori trusted. In 1999 Manu was also selling newspapers in the street but he helped (only in *O Espaço*) the work of preparing the newspapers for selling after the printing of the pages and was in charge of part of the interface between the agency and the *ardinas*: he distributed the newspapers to them, received the payment, and remained in the Square of the city during the selling (such as Disidori did in 1998). His main responsibility was in relation to the *ardinas*. This means that immediately after the time to start distributing the newspapers he left the work in the agency and assumed his role in connection with the *ardinas*. His first action was to write down a table with four columns: Taken, Sold, Left and Paid[7] and each row was allocated to each *ardina*. This was his tool for control of the whole process and Manu made this record visible to all those who wanted to see it during the day. He updated the records in the case that some of the *ardinas* took more newspapers to sell, writing down again in a different row the name and the number of newspapers taken. This was also his record for the final account at the end of the day. Then Manu organised the money and the newspapers not sold in order to make the final account with Disidori. Disidori usually went to Manu's home at the end of the day to collect the money and the newspapers left and to organise the activity for the next day.

The *ardinas* recognised authority in Manu but they kept considering him essentially as an *ardina* and they had the perception that he had not substituted Disidori ('*he is just helping him*' as Kaka said) and that he was not a member of the staff of the newspaper agency. For them, Manu had a special status in respect to Disidori and so indirectly to the institution.

The organisation of the distribution of the newspapers was different in 1999. The distribution of *O Tempo* took place on Friday at the agency of the newspaper away from the centre of the city. There was the intention of having *O Espaço* on sale on Thursday (almost every time this was not achieved) and the newspapers were delivered to the *ardinas* in a building (close to the Square) where the newspaper was printed and organised for the distribution. *O Tempo* was delivered to the *ardinas* by the treasurer of the agency. Before having the newspapers delivered to the *ardinas* the priority was to the people who took the newspapers to other islands and to the shops where now they were also sold. Only after that the *ardinas* received the newspapers and they had two options: Taking or Buying, according to their own words[8]. The agency of *O Tempo* defined a priority for the *ardinas* who buy the newspapers against those who just take them for selling. The decision on who receives and how many newspapers are delivered was a responsibility of the treasurer (sometimes with

---

[7] This way of organizing the records was not learned from Disidori. In fact Manu told us that he copied this process from a boy who did the work before Disidori.

[8] Buying means that the *ardina* should have enough money to buy the newspapers at the moment he receives them; all newspapers left after the selling are not returned to the agency and consequently not refunded. Taking refers to the process already in practice in 1998.

the interference of the director) but no explanation was given about the reasons behind those decisions. We could notice that the *ardinas* from S. Martinho had to buy the newspapers. In general when one of the *ardinas* did not pay back the agency for the newspapers taken, he had to come to the regime of buying instead of taking. The *ardinas* showed up at the agency of *O Tempo* in small groups during the day. At the same time there was now a frequent danger, the presence of pirates[9] who took the money from the *ardinas* and this made the staying at the door of the agency much more dangerous for them.

As far as the agency had newspapers to distribute they were given to the *ardinas* who showed up at the agency and in the quantity that they wanted. This meant that no newspapers were kept for those *ardinas* who were in school or that had other activities. The uncertainties about the number of *ardinas* that showed up to take or buy newspapers led to a situation were *O Tempo* was still dependent on them. In fact, the practice of buying the newspaper in a shop was absent in customers and we could even identify a certain resistance to this way of getting the newspaper. On the other side, the number of *ardinas* who assumed a commitment of exclusiveness to *O Tempo* was very low (only 4 out of 32 did have this commitment).

In *O Espaço* the person responsible for printing, organising and distributing the newspapers was Disidori (the same person who was responsible for the distribution of *O Tempo* among the *ardinas* in 1998). At the beginning, all the *ardinas* who sold *O Tempo* in 1998 showed up at *O Espaço* to sell this new newspaper (including all those from S. Martinho). While the two groups of *ardinas* (from Praia and from S. Martinho) were in the practice of selling newspapers, the distribution was decided by Disidori but was executed by himself and by Manu. So, Manu received a certain number of newspapers that he distributed among the *ardinas* from Praia and Disidori distributed the newspapers among the *ardinas* from S. Martinho. In March 1999, the number of *ardinas* from S. Martinho started to decrease and only Manu executed the distribution. However, Disidori decided if the newspapers were distributed all at once or if there were two phases for the distribution. For example, it was common that a number of newspapers were kept for the *ardinas* who were in school and because of that only come to the agency in the afternoon. Once the distribution was finished, Manu came to the Square carrying a certain number of newspapers to give to some *ardina* who was late. He seated himself close to those people who sell goods in the place and there he was during the rest of the day orientating, controlling and in general showing that he was available to receive the final payment just after the selling. The new role of Manu demanded a different approach to the practice itself and to his *ardinas'* peers and, at the same time, a different point of view towards the calculations involved in the situation. All this necessitated a new learning process for him. His previous learning happened (naturally) within his participation as an ordinary *ardina* and was not specially designed for his new role. Participating in different moments of the history of the practice (which organisation was transparent for the *ardinas* about several issues of the controlling process) enabled him to pick up

[9] *Piratas* (pirates) is the word in Praia to refer to children who make small robbery in the streets.

useful elements for that learning. This was the case, for instance, in the table to control the selling process. Now, he was participating in a slightly different practice (the controlling one) in a way that we can classify as peripherally legitimate[10] (towards the institution). This enabled him to use resources from what he previously observed in Disidori's behaviour but also from what he experienced himself as an *ardina*. For example, it was possible to observe some similarities with Disidori's behaviour in his way of dealing with the *ardinas* but also some differences.

Because there were a greater number of *ardinas*, now they had to circulate more around the city and go further but they kept having their own places for **selling**. It is important to reiterate that just a few *ardinas* followed the rules expressed in December 1998 by the administration of the newspaper *O Tempo* (to avoid selling both newspapers at the same time). But even those *ardinas* were concentrated close to Manu or in small groups. For example, Manu (with responsibilities in *O Espaço*) was selling *O Tempo* every time he could and Bétu (who was in *O Tempo*) also found a way of selling some newspapers from *O Espaço*. This means that all the *ardinas* were sharing the same places for selling and maintained the relationships among them. They helped each other with the change (exchanging coins and bills), they stayed talking after the end of the selling or in the pauses and they even called for someone who had a certain newspaper if a customer asked for it. Most of the *ardinas* managed to have both newspapers to sell and they arranged this going and getting both newspapers at the agencies or exchanging newspapers among them. In their opinion, the fact that they had the two newspapers for selling was something that facilitated selling both of them. The price of the newspaper for the customer was the same (100 escudos) and the profit per newspaper was 20 escudos in both cases.

The principle accepted by the *ardinas* was that if they took a certain number of newspapers for selling, at the end of selling them they should pay back 80 escudos for each one sold together with the newspapers left. For *O Espaço* the **payment** was carried out in the Square of the city as soon as the *ardina* finished selling his newspapers. There was no collective final moment for this operation. However, because this was the place of selling for most *ardinas* (and the place where most people working at the city were passing by) it was common to see a group of *ardinas* together, some of them selling, some paying and some just observing. For *O Tempo* theoretically the payment was done after selling and at the agency. But we observed that several *ardinas* preferred to make their way to the agency in groups because it was a long way and they were carrying a significant amount of money. So, they stayed at the Square after finishing selling waiting for other *ardinas*. The Square became a place to be naturally together and this gave rise to the same kind of environment that was observed in 1998.

There was in 1999 a quite different way of integration of the newcomers. In *O Tempo*, as was mentioned, the newspapers were distributed directly to all the *ardinas*

---

[10] Legitimate peripheral participation is proposed by Lave and Wenger as a 'descriptor of engagement in social practice that entails learning as an integral constitutent' (1991, pg. 35).

(except for the three boys protected by Bétu). There was no kind of explicit attitude to a newcomer. But it was possible to identify that a system of integration was in practice. The boys who wanted to start selling newspapers tried to get into this practice through a friend or a neighbour who managed to have a number of newspapers given to the newcomer and assumed a kind of orientation of his first steps. It was also this boy who went to the agency to pay for the newspapers sold by both and not the newcomer. Apparently the system was functioning as previously. However, the price that the old-timer received from the newcomer was different. During the learning phase the newcomer received only 15 escudos for each newspaper sold (instead of 20 escudos). This process was taking place out of the control of the administration of the newspapers but it was a short period because very soon the newcomer got in touch with other *ardinas* and understood the difference in the income. Given that he felt comfortable in the practice he started going to receive directly the newspaper from the agency.

In the case of *O Espaço* it was possible to observe a slightly different process. The newcomers arrived as friends of old-timers but it was Disidori who had the last word about the acceptance and he always tried to have some references about the newcomer. After that, some short indications were given to the newcomer (such as the price of the newspaper and where to pay) and he started selling. There was no one institutionally responsible for the newcomer but informally the *ardina* who brought the newcomer usually assumed a role in accompanying him. The *ardinas* believed that in this process the most important was the protection and to show the good places for selling. All the other aspects were considered as naturally accessible and easy to learn. But even with differences in 1999 it was clear that some of the fundamental moments of the process of learning to be an *ardina* were lived in the Square both when they were selling together, or paying or listening to the stories of the happenings of the day.

With this description we intend to enable the reader to have a sense of the historical transformations that arrived in the context of the *ardinas'* practice, particularly in what concerns their participation and their learning. Doing so, we are now in position to go deeper in some issues that emerged from our analysis focused on transitions experienced by the youngsters. The main goal of the next section is to support the claim that the *ardinas'* mathematical point of view and their 'mathematical' problems are strongly situated in that practice and follow their growing in competence of being an *ardina*.

2. COMPETENCE WITHIN THE PRACTICE AND MATHEMATICS

Julien (1997) argues for a different way of looking at the relation between competence and the taxonomic features. The traditional way of dealing with knowledge (in particular in education) has led us to accept, almost without being contested, that in order to be competent in any domain one first needs to 'acquire' a certain amount of objects of knowledge. This idea also seems to be behind the 'natural' acceptance of the hierarchical order of the mathematical learning process. It is very often said that in order to be able to deal with (or use) certain mathematical objects or ideas, one needs to learn or understand a lot of other previous 'fundamental' mathematical concepts. Only after learning in school a certain amount of pieces of mathematical knowledge the students will be able to apply them in out-of-school situations. However, Julien (1997) points out that competence is not built from facts but in fact it is the other way around, 'the basis for the production of facts' (pg. 270) and states that 'taxonomic features are not the building blocks of competence; they are a discursive product of such competence' (pg. 270). He builds his approach both from the situated cognition and the connectionist theories.

For the purpose of this chapter, we will focus on some of his ideas trying to enlighten the connections to some particular features of the situated perspective of learning (Lave & Wenger, 1991). We will present and analyse episodes of *ardinas'* life that we believe can give support to the idea that their mathematical knowledge is found in the practice in which they participate. For instance, it is the case of the context, the structuring resources, the habits that gets them through the daily activities in which they are involved and in their discursive practices. We will focus on particular cases, describing some details of the situations but not forgetting their location as moments of the *ardinas'* practice described early on.

The main point is to show how mathematical knowledge emerges from the competence of being an *ardina*. For this purpose we will focus on two instances of transition: the first one draws on the research situation and not only the *ardinas'* practice, that is the transition experience arises from the situation of the *ardina* acting as an informant to a researcher; the second is one of the practice itself or in other words, the transition is closely related to what happen within the practice and very much connected with the changes that emerged from the historical transformation of the social and economical world that surround this practice as well as from the integration of new members in the community of practice.

2.1. Calculating within the Practice and Talking About It

Certain aspects of the *ardinas'* practice require dealing with numbers and calculations. However, as was already mentioned, a great number of *ardinas* (particularly the old-timers) had very low school qualifications and had left school some years previously. Very early on we were faced with the issue of identifying relations between the *ardinas*' participation in their practice and the mathematical knowledge that they developed. During this search several opportunities arose to clarify the relation that *ardinas* established between that knowledge and school mathematical facts. Since the main concern of this research is to gain as much access as possible to the genuine aspects that are typical of the practice in study, we avoided experimental situations and tried to integrate the practice of dialogue (between the observer and the *ardinas*) in what they could accept as true interest in understanding their practice. In Cabo Verde children and youths are easy-going about contact with strangers and usually curious about them too. However, the observer's characteristics – a woman (as were most teachers), speaking Portuguese (the school tongue) more fluently than Creole – might hinder her access to these youths' practice and knowledge. In other words, there was a risk that they might interpret her need to ask questions (many regarding mathematical aspects) more in terms of school evaluation than as a true interest on her behalf in understanding what they knew and she did not.

The dialogues we analysed always took place in the presence of the observer and most of them were brought on by her while talking with the *ardinas* during breaks in their sale or while waiting for the distribution. Within the *ardinas'* practice there were not many moments where the 'natural' dialogue between *ardinas* (and with Disidori or Manu) involved reasoning that included discursive elements that are mathematics-related. We figured that this kind of reasoning would happen in payment situations or during the breaks in sales, at which time they verified the money they had. But, in fact, we rarely found this over a period of almost six months for data collection. Usually they checked (the money and newspapers) very discreetly, as if in a private way even though others were physically present. That is, they counted the newspapers and money in silence, without talking about it to each other. It was a painful experience if they happened to find that the money did not correspond to the number of sold newspapers, as if this revealed their inexperience (or lack of skills) regarding the practice. So during these moments they were somewhat reserved in the presence of the others and did not share their doubts.

Similarly, during payment, if any *ardina* faced a situation of disparity between what he expected to pay and what he was asked to pay, his behaviour was discreet and non-argumentative. He would try to delay payment, showing signs of doubt such as looking once again at the calculator screen or recounting the money. This way Disidori or Manu were forced to do their calculations again, paying more attention to the numbers turning up on the calculator. Some *ardinas* did not even hint at it, simply accepting that the value on the screen of the calculator was correct and handing over the money. At the end they would walk away with their heads low, feeling sad but without saying a word on the subject. If the observer approached them and talked to them about the matter they would often explain what had happened as proof that they had lost money during the sale, or even they had given too much change back, or someone had stolen some newspapers, or else they had not received the correct number of newspapers from the start. Since Disidori's and Manu's calculation processes were completely transparent, they watched the numbers Disidori put in the calculator (*'to see if he's doing the sum well'*) and assumed that the calculator never made mistakes and so the result was never defied.

Therefore there were normally few discursive opportunities during this practice as far as aspects that we usually identify as mathematical are concerned, yielding difficult access to these youths' own way of thinking mathematically in their *ardina* practice. However, in some moments this access was possible and we found certain interesting aspects such as their calculation strategies, the supporting elements for these calculations and the difference between thinking and making this thought visible and understandable for others. Situations of conversation with the observer, seen as a person outside the practice but interested in it, also offered the opportunity to approach other parts of the *ardinas'* lives and to talk about their school practice in a natural form, for example.

We are going to look into three aspects we gained access to by means of the analysis of several situations, that we think may help us to reflect upon the relation between the competency the *ardinas* develop during their practice and the mathematics that is present in this practice. First we will discuss the supporting elements for calculations available in their practice which are typical of it and arise in the *ardinas'* calculations processes, secondly the differences between actually calculating or talking about this calculation, and finally the existence of traces of the school discourse and practice.

2.1.1. Supporting elements of the calculation

When we asked the *ardinas* something about the processes they used to make sums while selling, most of them, particularly the newcomers, illustrated their explanation with the act of paying. For example, in 1998 Konka (an 8th grade student) explained how he thought about the money he made in the following manner: '50 newspapers is 5000 escudos, then you take money off to pay Disidori and keep the rest'. The value they had to pay for each newspaper (87.5 escudos) was too difficult for them to actually use it in their calculations. However, after a while in the sales business, almost all of them began to handle the value of their gain (12.5 escudos) adequately. The old-timers and some of the more school qualified ones could sometimes explain how they foresaw the value of the money they should pay or make in hypothetical situations that were put before them.

For example, talking with Djeps (one of the most inexperienced *ardinas* in 1998) at several moments during data collection in 1998 it was possible to identify a sequence of different forms.

Epis. 1

One day in the second week in Cabo Verde (while waiting for the distribution with some *ardinas*) the observer asks Djeps if he knows how much he would earn with the selling. He begins to explain spontaneously what should be the gain for 5 newspapers[11]

Djeps – One is 12.5 escudos; 2 is 25;... 4 is 50;... 5 is... 62.5 escudos.
Obs – And about 8 newspapers?
Djeps – (*after some seconds*) 100 escudos.
Obs – How did you think?
Djeps – Two is 25; two more is 50,... are four; plus four are 100 escudos.

[11] The dialogues between the observer and the *ardinas* were in most cases a mixture of Portuguese and Creole. Generally the *ardinas* expressed themselves in Portuguese with difficulties and often they jumped very fast to Creole. The translation from Creole or Portuguese into English presented in this report certainly transforms the original meanings as in every translation process. We tried to keep the translation of the *ardinas'* utterances as close to the meanings as possible with sacrifice of the elegance of the wording in English.

Some weeks later he described, for instance, how he calculated the profit on 40 newspapers as follows:

	8	16	32	+8	40
are	100	200	400	+100	500 escudos

It took a while to get to this point but in 1998 we could find this form of calculating the profit (in action and in their explanations) in several other *ardinas*, whether they went to school or not. In other words, in order to calculate the profit they would often use the corresponding value of 8 newspapers (100 escudos) to support this calculation. Besides one newspaper costing this much, this value was also one of the notes (of Cabo Verde's monetary system) that were most used in selling newspapers. When they went to pay, most of the notes they had were usually 100 escudos and 200 escudos notes, as well as some 50 escudos coins.

Other interesting moments happened in 1998 with Kodé (a competent *ardina* from S. Martinho who finished the 6th grade two years before).

Epis. 2a

Kodé passes near the observer with 12 newspapers and tells her that he already sold 63 (he received 75). She asks him how much did he earn and after some seconds he answers:

Kodé – 887.5 escudos.
Obs – How did you do?
Kodé – 50 plus... 725 plus 13, is 162.5 escudos.
Obs – How did you do for the thirteen?
Kodé – Four...50; 8... 100; 12... 150;... 162.5 escudos.

We can see here that the number of newspapers (63) is split into 50 and 13. (He has added an extra 100 escudos, for the initial 50 newspapers, without realising his mistake.) The number 50 (just like 25 and 75) is frequently used as a base for calculation, since these are the most frequent quantities (in 1998) of newspapers they receive for sale[12]. On the other hand, 13 is sub-divided and the values 4 and 8 emerge (50 escudos and 100 escudos, respectively). This addition process which starts with one of the base numbers of received newspapers is a common case (both in 1998 and in 1999). The case of using the value of the profit of 4 and 8 newspapers is only verified in 1998 (12.5 escudos $\times$ 8 = 100 escudos).

Epis. 2b

One week later, waiting with Kodé for the distribution, the observer presented him with an hypothetical situation: he receives 75 newspapers and after selling them all he will receive 20 more. She asks if he would be able to know the total amount of money earned but using the

[12] In the case of the younger *ardinas* (like Djeps at the start) there is also the 5 or 15 which are quantities handed over to the younger ones when they begin to participate and are learning from an old-timer.

calculator. He begins to do 75 + 20 and multiplies by 87.5 escudos and says 'Is to pay'. After he divides by 87.5 escudos and in the screen appears 95. He stops and can't do anything else. Then she asked if he is able to do it without the calculator.

Kodé – 1250 is 100,... less 62.5 escudos, that is 5 [newspapers].
Obs – How do you know that 5 is 62.5 escudos?
Kodé – Twelve and half more 50... as 25 is 2, 50 is 4.

It is now possible to see that Kodé calculates his profit through a process of subtraction from the value of 100 newspapers. This form also arises but only in more experienced *ardinas* or in those with more school qualifications. Indeed, this *ardina* was one of those who had finished the 6th grade but was referred to (by his brother and by Disidori) as having been a very good student and, as he himself said, enjoying mathematics. The difficulty he showed in using the calculator (which is not used at school) shows how this instrument, despite being present in the *ardinas'* everyday life, is actually not part of the artifacts of their practice. We only observed the use of a calculator by some of the old-timers and even then it was just to calculate the value they had to pay. The calculator is an artifact of Disidori and Manu's practice when controlling the *ardinas'* payments, but it is not one of the artifacts of the *ardinas'* practice.

Epis. 2c

Following the talking with Kodé, the observer asks him how much should be the earning for 95 newspapers if they earn 15 escudos for each one. He begins saying he did not know.

Obs – You know how much is 100 newspapers?
Kodé – 1500.
Obs – And five?
Kodé – 75... if it is 95 it would be 1425.
Obs – And if it was 20 escudos?
Kodé – Gives 2000 escudos.
Obs – All 100? (he nods yes) and the 95?
Kodé – That will give 1900 escudos.
Obs – What is easier, with 15 escudos or with 20 escudos?
Kodé – With 20 escudos
Obs – Why?
Kodé – Cause 5 newspaper gives an exact number, 100 escudos.

It is interesting to look closely at what happens in this dialogue. First a situation is considered, which is hypothetical but involves a rather 'uncomfortable' number for the mental calculation that his practice requires. Indeed, it is not easy to quickly reach a multiple of 15 escudos that is associated either with the newspaper's value or with some of the coins or notes he used. Maybe this is why his first reaction was to deny knowledge of a process for solving this situation. After suggesting the process that he himself used in the previous case, he actually manages to calculate the profit both in the case of 15 escudos and in the case of 20 escudos. One aspect that deserves special attention is the way Kodé justifies the greater facility in calculating with 20 '*5... it gives an exact number, 100*'. This is his way of highlighting the aspect (already mentioned) concerning the search for values that coincide with elements of the monetary system or the price of the newspaper. This was the reason why multiples of 8 were used to deal with the value 12.5 escudos in 1998 and multiples of 5 used in 1999 to deal with 20 escudos.

The calculation of doubles and successive additions are other common aspects that most *ardinas* use in their explanations of different calculations, when the proposed situations involve unexpected or unfamiliar values. For example, the following description is used to explain a calculation with 15 escudos: '*2 is 30, 4–60, 6–90, 8 is...*'.

In 1999 it was easier to find *ardinas*, even amidst the newcomers, who very early on could perform calculations both in relation to their profit (20 escudos) and in relation to the value they paid for the newspapers (80 escudos). Also, they more frequently used the reference of 25 (newspapers) than others (50 and 75) particularly among the *ardinas* who began to sell in 1999. At the *O Espaço's* distribution most of the *ardinas* received 25 newspapers; we can understand that those *ardinas* were more familiar with the situation of receiving 25 newspapers than any other kind of number.

In summary we found a habitus on the *ardinas'* practice in dealing with calculations in both periods. That is, all the *ardinas* arrived at using certain elements of the context to support the organisation of their calculations:

(i) the sequences of mental calculation were structured with supporting elements that concerned values which were strongly related to the elements of the monetary system and to the price of the newspaper;
(ii) the numbers such as 25, 50 and 75 were used as reference for thinking (the 100 was even more rare in 1999 than in 1998;
(iii) the calculations began with the values they would have to pay (the newcomers) or the values they made, rarely starting with the money at which they sold the newspapers ($100 \times x$);
(iv) adding processes were used more often than subtracting processes.

2.1.2. Internal vs. external conversation or framing vs. re-framing

On this topic we shall turn our attention to aspects of the practice that gain special visibility in different discourses in situations such as 'thinking aloud', which is typical of the *ardinas'* practice, or explaining what goes on in the practice to an outsider (the observer). In other words, these are moments when thinking is externalised in a manner that resembles Restivo's (1998) 'thinking as internal conversation', or that which we considered to be an external conversation and is usually found when the *ardina* explains how he thought. This is almost as if in one of the situations the *ardinas* were more 'genuinely' in the framing of their practice and in the other they were in a situation of re-framing of their practice, for outsiders (Goffman, 1991, 1974). This way not only are we able to understand relevant aspects of the practice from the insiders' point of view, but we also see what *ardinas* use to make it understandable for others.

Two specific examples (Ntóni and Manitu) will be presented and discussed, showing some of the aspects that have been highlighted as supporting elements of calculation.

Epis 3a

During the selling in 1999, Ntóni sits for a moment on a bench (with another *ardina* still selling near by). He puts his bag on the floor and the newspapers he has on his lap. The observer looks at him and understands that he is counting the newspapers he has in the bag. She feels that he has some problem, so she goes to sit near him. He is trying to figure out if he is missing some money or some newspapers.

Obs – Did you lend newspaper to someone? (*he nods no*) No? What then?

He counts again the newspapers within the bag and after the ones he has on the lap. Then he begins counting the money he has (coins and bills).

Obs – How many newspapers do you have there? (*the observer put her hand on the newspapers on his lap*).
Ntóni – 5, 10, twenty (*he knocked on the newspapers with the fist*)
Obs – Do you have 20 newspapers, all together? How many did you take?
Ntóni – 38.
Obs–38?! Then how much money should you have there?

He recounts the money (the bills), doesn't answer the observer and during the counting looks around as he is thinking.

Obs – So? 38, you have here… 20?
Ntóni – 35 (assertively)
Obs – 35 or38?

He doesn't care about her and starts again playing with the bills and coins.

Ntóni – *Twenty (he is thinking about the nineteen newspapers he carries at the moment and the correspondence to the coin of 100 escudos; after he puts together the 500 escudos bills)*… 25, 30, 35,… missing 3.

This episode shows that Ntóni's actions and words reveal a fundamental concern to make sense of the unexpected situation that arose. That is, to certify whether he has the money and the newspapers unsold corresponding to the 38 newspapers that he thought he had received (and had told so to the observer some time before) or whether he only has 35. His discourse was audible, though not entirely directed to the observer, almost as if he were talking to himself. The structure of the talk shows that he was clearly not worried about what the observer understood. Although he seems to maintain a chain of interaction rituals that would suggest a dialogue (that is, a conversation in which the speech of one projects itself in the speech of the other) it is not, in fact, a true dialogue content-wise. The observer's posture remains that of an outsider trying to make sense of the *ardinas'* actions, behaviours and thinking. Ntóni, on the other hand, now shows that he is not meeting that need of the observer (though he does not refuse it completely). The framing of his interaction is rather that of an internal conversation, of a search for meaning regarding a situation in his practice as an *ardina* where something unexpected has happened (though typical of this practice and in the scope of his skills). This will be actually a problem at the time to pay and brings problems to the satisfaction of his need to participate in this practice.

Let us go into more detail about this episode. In reply to the question about the number of newspapers he has on his lap he first says 5 (that are indeed there), seeming to have heard the question and knowing which newspapers the observer is referring to. However, he then goes on to the number 20 (instead of 19 that are the newspapers he has not sold yet). In these 20 he is already including, besides the 19, the 100 escudos coin he has in his pocket (in the money he counted just a few moments before). When the observer questions him once again, this time about how much money he must have in his pocket in relation to the newspapers he says he has on him, his actions begin by counting the money but the sequence of the words that follow between both of them shows that his answers have nothing to do with what he is being asked. In his last utterance Ntóni verbalises a sequence of numbers indicating the counting of newspapers while he handles the money (bills), ending at number 35 and stating that 3 newspapers are really missing. In this check he counts newspapers naming each note to be a given number of newspapers, thus using a counting strategy more than a calculation strategy.

Epis. 3b

Obs – Thirty five, it misses 3 newspapers, so here you have money of 35 newspapers? (*he nods no*) No!... Here you have money of... 500, (I start counting the money he shows me)... one thousand and 500, one thousand and 600. One thousand and 600. How many newspapers is this?

Ntóni – 16.

Obs – 16 newspapers and now those more... (*Ntóni count again those he has on his lap*).

Ntóni – 19, here there are 14 (*he points to his bag*) and here are 5 (*he points to his lap*).

Obs – 16 plus 19 how many newspaper is it? Yes, there is 35, there are 3 newspapers missing, how did you lose money? (*he puts the 5 newspapers in the bag and get up*) Do you think that you gave newspapers to another boy? (*he nods no*) Did you count the newspapers when you received them? (*he nods yes*) It was correct?!

Ntóni–Yes (*he carries the bag and gets away from me alone*).

In this part of the episode Ntóni's speeches and actions now seem to be a compromise between the framing of the practice (checking once again) and the talk with the observer (answering her implicit request to gain access to what he is thinking). From this point on he actually answers each one of the observer's questions in a genuine external conversation which is more explicit and which has a sequence of speeches and actions that seems to show some concern in having the other person understand what he thinks (*19, here there's 14 and here there's 5*). Curiously, when the observer counts the money saying '*one thousand 500, one thousand 600. One thousand 600*', she shows she is actually not in the framing of the *ardinas'* practice at Ntóni's level, she does not count newspapers while handling the money, but adopts her usual way of being, which allows her to make sense of the situation more rapidly. On the other hand, we should emphasise Ntóni's final attitude, distancing himself after having given the observer the basic information she requested regarding justification for the happening. In a way it reveals his concern to get back to selling and recover the time lost during this check, but this isolation also shows that he is sad and also a little 'ashamed' of not yet being a very competent *ardina* (he is still deceived by others).

Let us now take a look at another episode where we can find what we understand as an example of thinking as internal conversation and where we can go further in the analysis. This one took place with Manitu – a slightly older (age-wise and in terms of sales experience) and more competent *ardina* than Ntóni. Their schooling was very similar, Ntóni still attending the 6th grade and Manitu having finished the same grade in 1998.

Epis 4a

One day in 1999 Manitu received 75 newspapers, he and the observer are talking in the afternoon during some time when he is resting a little. He was walking in her direction, dealing with his money and organising it, and stopped near the bench where she was. It seemed acceptable to talk with him about the checking he said to be usual to do during the selling phase. The observer was trying to understand how he really does that checking…

Obs – And how do you know if the money and newspapers are correct?
Manitu – Count money and count newspapers, then in my head (*he points to his head*)
Obs – So, how many newspapers did you already sell?
Manitu–Sixty three (*as he thinks*)
Obs – 63, how much money do you have there?
Manitu – Hum?
Obs – How much should you have there? For the 63.
Manitu–800… and 60… 960… I should… 5 *conto*[13] and… seven hundred and…40… so,… 960… no, from 960… and more… 5 *conto* and… 400 no,… 5 *conto* and 40.

It is worth noticing how this *ardina* describes the checking process, fundamentally through counting physical objects (money and newspapers) and then mentally. Next, the observer explicitly refers to the number of newspapers that Manitu has already sold (63) and tries to focus his attention on the money he was handling before (the money from selling these newspapers). From here we can see that she foresaw the existence of a checking process by which the *ardina* would try to verify whether the full money corresponded to the number of newspapers sold. However, Manitu verbalised an unexpected sequence of numbers. He is not explaining how he thought. He is actually thinking out loud, as if this interaction's framing were more one of thinking than of dialoguing with the other person, somewhere very close to the notion of thinking as internal conversation.

Let us go into some detail regarding the sequence he presents so that we understand his line of thought. He starts by thinking not about the value of the newspapers

[13] *Conto* is a Portuguese word that stands for one thousand escudos. It is used by some of the *ardinas* to refer to one thousand escudos from Cabo Verde.

he has already sold (63) but about the value of those he has with him (12 newspapers still unsold) and what he will have to pay for these (12 × 80 escudos). Then he seems to be trying mentally to calculate the difference between the value to be paid for the newspapers he received (6000 escudos = 75 × 80 escudos) and these 960 escudos. First he refers to 5 *conto* (if from 6 thousand escudos I subtract a number very close to one thousand naturally I will reach a number close to 5 *conto*) but then he mentions 740 which, amidst hesitations, is changed to 400 and finally into 40. In other words, he seems to be convinced there must be a four in the final value. If his calculating method were purely oral (without the interference of school algorithms) it would be more likely that he would immediately add the extra 40 escudos he had withdrawn when he rounded 960 escudos to 1000 escudos. In that case it would have been quicker to reach the result of 5040 escudos. However, the numbers 740 and 400 allow us to wonder about the possibility of having used another form or of being influenced by certain habits related to the written form of subtracting in school. If he was visualising the algorithm (6000 – 960) this confusion concerning the location of the 4 would be acceptable and therefore he would hesitate between the four hundred and the forty.

Although the issue that was raised concerned all the money corresponding to the 63 sold newspapers, the calculations that were performed show he was calculating the value he would have to pay for those 63 newspapers. Let us continue the episode.

Epis 4b

Obs – Five conto and 40... Why? The money for you paying or the money you got from selling for the moment?...
Manitu – No, for paying...
Obs – Ah! For paying.... and for the money all together, paying and your profit...
Manitu – All?
Obs – You still have 12 newspapers... how many did you sell? Is not your profit, is not for paying, what total value of... 63 newspapers?

In this dialogue Manitu confirms his concern in calculating the value he must pay, which leads the observer to think that perhaps her question was not very clear. So she tries to reformulate it and clarify that she would like him to think about the full money involved in the sale of 63 newspapers.

Epis 4c

Manitu – 63 newspapers? Four... no, 50 is four conto...
Obs – 50 is four conto?
Manitu – Yes... One conto and three hundred,.., no 5 conto and three hundred.
Obs – For paying?...
Manitu – No for...
Obs – 50 newspapers is 4 conto? (*he says yes*) For paying... to Anriki?!...
Manitu – Yes for paying, then the profit...
Obs – Yes, then the profit...

It would be interesting to go deeper into this moment of dialogue where he actually seems to be replying to the observer. Yet he goes in a different direction from that shown before. His concern in answering the observer's needs is shown from the beginning of his explanation in the structure of the phrase '50 is four *conto*'. As we referred earlier, it is unusual for *ardinas* to talk about their calculations and some of them use a phrase structured like this to clearly express their will to explain what they think instead of simply answering the observer's explicit questions. Now he begins by calculating the 50 newspapers and not the 12 he still has to sell, that is, he now uses a new arrangement of the 63 newspapers (50 + 13). Although he still thinks first in terms of the value to be paid for the 50 newspapers (4 thousand escudos) he comes up with a value (one thousand three hundred) that corresponds to the sale of the remaining 13 newspapers. This leads to a value that corresponds neither to what he has to pay nor to the value of the 63 newspapers' sale.

In the first part of the episode (his internal conversation) Manitu did not get lost in the sequence of calculations. However in the second (a possible attempt to re-frame his thinking) he confounds his main need as an *ardina* (to pay correctly) with the need arising from interaction with the observer (to explain how he thinks). This interaction involves a different and unusual discourse in this context (that of talking about mathematical facts within the practice), which disturbs his line of thought. This occurs due to this transition between two practices – that of an *ardina* and that of an informant on *ardinas*' mathematical knowledge – that call for different competence. It is possible to say that upon performing typical calculations of his *ardina* practice, Manitu found and used points of support in it that allowed him to keep a solid line of thought. Doing so, he was able to respond adequately to his problem (to check if he has the money to pay for the sold newspapers). Yet when he is faced with the need to speak out this thought he is taken away from his *ardina* practice and it is as if loses those secure points of support, living another practice in which he is not yet confident that he is sufficiently competent. In this practice, not only is the problem defined in a different way, but it can be understood entirely as another problem – not that of an *ardina* in his daily life, but that of an observer trying to understand (and learn with) *ardinas'* practice.

We think that it is really difficult (for an outsider as it is the case of an observer/researcher) to have deep access to the mathematical thinking process characteristic of the *ardinas'* practice, as the actual language-game of the practice does not include discussions about the mathematical facts we know to be present in that practice. As if although mathematical thinking is present within the practice the mathematical arguing was not 'allowed' to happen within that practice. So we almost only have access to the external conversation with which the *ardinas* are re-fraiming their thinking for the observer (with a different structure from their thinking as internal conversation). Even though the few moments where it was possible to analyse both external and internal conversations showed some key features of *ardinas* thinking about calculations:

(i) when they have both newspapers and money to check out, they 'transform' money into newspapers (500 escudos are 5 newspapers);
(ii) they were more concerned with checking if they have enough money (80 escudos per each) to pay back the newspapers sold then the total amount of

money (100 escudos per each) they should have in their pocket corresponding to the number of newspapers sold;

(iii) when they need to re-frame their thinking they try to adjust it to what they think are the needs of the observer and doing so they can make more mistakes than when they do their calculations within their practice.

2.1.3. Traces of the school mathematics discourse

Within the vaster project of which this paper is part, we found that all *ardinas* (from the least school qualified to the most qualified as well as those presently at school) ended up using similar calculation strategies, with the same supporting points and scarcely using school mathematics strategies. However, we observed that at certain moments the *ardinas* seemed to use some mathematical knowledge in a school-related form. This was the case of Manitu (in the last episode), for example, but these moments occurred more often in situations born in the interaction with the observer.

At this point we shall analytically describe some episodes that reveal the several forms these 'school traces' had.

Epis 5

In 1999, around a tree in the Square where a group of *ardinas* is paying to *O Espaço* newspapers they sold. Ntóni is near by but carrying some newspapers *O Tempo* for selling. He observes the colleagues but he is still paying attention to the movement around him trying to see some customer. The observer is near him.

Obs (to Ntóni) – How many newspapers did you take today?
Ntóni – 38
Obs – Did you take 38? Why?
Ntóni – There were no more.
Obs – There were no more?! Anriki is the one who says how many you should take? (*he nods yes*) Ah! So, now you will sell all of those until the end of the day? (*he nods yes*) And how much is your profit for the 38? Do you know?
Ntóni – Seven hundred and sixty (*slowly, as he was thinking*)
Obs – 760? How did you think? Do it for me.
Ntóni – Multiply (*he keeps alert about what is going around him*)
Obs – How did you think?
Ntóni–In my head...
Obs – But do it, how do you do it in your head, how do you multiply in your head?
Ntóni–At 25 I earn 500; 10 is... 10 is ..., 15 is 300 escudos; 3 is ...60 escudos; 360,... I earn 860.

At that moment another *ardina* arrived to show him something and Ntóni went away with him.

Once more (as we noted in the last point) the *ardina* makes no mistake when he spontaneously performs calculations about the real values at issue. But this does not happen when he has to explain his calculating process to the observer, at a time when he is still concerned with the sale and his attention is turned to this. When he explains how he reached the value of 760 we find again the supporting points that are usual in this practice when: (i) the 25 (the number of newspapers he receives every week at *O Espaço)*; (ii) a sequence of additions whose parts are multiples of 5, that is, of 100 escudos. However, there is one slight detail that stands out in this dialogue. Ntóni's first attempt to explain how he reached 760 is done through a word (multiply) that does not belong to the framing of an *ardina's* practice, but rather to that of school.

As showed in the last episode, in which Ntóni includes words from the school discourse in the dialogue with the observer, this often happened with other *ardinas*, even with those for whom school was more of a memory than anything else. When trying to explain to the observer the procedure for calculation (and particularly those who had more difficulty in expressing themselves in Portuguese), sometimes the *ardinas* used words that seemed to show some relation to their thinking, although these words were not always used correctly. The following episode illustrates this.

Epis. 6

The observer is talking, in 1998, with Pitchiu (an *ardina* from S. Martinho who left school 4 years ago) before the distribution. She is with a colleague from Cabo Verde (Ana) who acted during the 3 first weeks as an interpreter between the observer and some *ardinas* who have more difficulty understanding and talking Portuguese. They were talking about the profit they do from selling and Pitchiu is saying that he usually sells 100 newspapers so he has a profit of 1250 escudos. She asked how much he would earn if the profit for each newspaper was 15 escudos. He needed some seconds of concentration but he said with no doubts that it would be 1500 escudos. The observer asked him to explain how he had thought. During several attempts he was doing, in creole, to the observer or to Ana some dialogues developed such as the following.

Ana – Think aloud, make the calculation aloud...
Pitchiu – I do in my head...
Ana – How much is it, your profit?
Pitchiu – One thousand and five hundred.
Ana – How did you do it?
Pitchiu – In head... I added, added...
Obs – How did you add?
Pitchiu – Added 15 times 50,... it gives,... gives... gives.... 750 escudos times 750 escudos...
Kodé (*an ardina, friend of him that was near them*) – No, he has done the calculation of 50 times 15.
Pitchiu – 50 times 15 gives 750... 750 plus 750 gives 1500
Obs – Ah! And 750 how did you think?
Pitchiu – Added from 1 to 50... 50 is half of 100. Added 15 times 50.

Here the verb 'to add' seems to be used more to adapt to the observer's discourse; it is not a word of the *ardinas'* practice, there the words they normally use are 'plus' and 'times'[14]. We may think that their difficulty simply stems from their foul

[14] Although those words are commonly used by most people, they are strongly related to the *ardinas'* practice for this is what they use when they describe what Disidori or Manu do with the calculator in order to tell them how much they should pay – for example, '*50 [newspapers] times 87.5 escudos*'.

Portuguese, but in other situations, where they explained other aspects of their life as *ardina*, they commonly turned to Creole to make themselves clear. They had already realized that the observer understood them well in Creole (actually one of the *ardinas* with whom she most talked almost always spoke in Creole). This led us to consider that talking about the calculation processes they use is not a part of the practice. This result is coherent with the observation and with the information gathered from several elements (*ardinas*, Disidori, Manu, former *ardinas*).

But the presence of the school influence was felt in more than scattered words in certain moments. We can see this in the following dialogue. It takes place in the Square, between Diku (in 1998) and the observer, during a break in the sales. This *ardina* was from S. Martinho and he had attended school until the previous year, having been a very good student and concluded the 6th grade.

Epis. 7

The observer is talking with Diku about the selling of the day before when he sold 35 newspapers.

Obs – How much was your profit?
Diku – Twenty five is 325 escudos plus 10 is 125 escudos.
Obs – When you were young did you think like that for your calculations?
Diku – No.
Obs – How was it, then?
Diku – I did with the calculator.
Obs – Did you learn with somebody?
Diku – No, was by myself…
Obs – How did you begin to think about it, did you remember?
Diku – 25 at 12.5 escudos gives 312.5 escudos;… 50… 25 is half of 50,… gives 312.5 escudos … 50 gives 625 escudos.
Obs – Did you already know that 8 is…?
Diku – 100 escudos
Obs – How did you begin to think about 8 newspaper? You never take 8 newspaper, do you? (*he nods no*)
Diku – 8 is 100 escudos because 2 gives,.., one is 12.5 escudos, 4 I'll earn 50 escudos and 4 is half of 8, 8 is 100 escudos, 16 is 200 escudos, 20 is 250 escudos.
Obs – 20? Why did you jump from 16 to 20?
Diku – Because 16 is 200 escudos, 4 more is 50 escudos, so 20 is 250 escudos.

In this episode Diku clearly explains his calculations by means of the same supporting points of the practice that we referred to earlier on (the 8, the 25 and 50 and the doubles). However, the structure of his explanation is somewhat different from what we usually find in the dialogue with *ardinas* who are less school qualified or attended school longer ago. When the observer asks the first question his answer is not the final value but the way of reaching it. Other *ardinas* would normally reply the former, making it necessary for the observer to make explicit the question of how they did it so that they would describe processes. It seems that this particular *ardina* acknowledged a sort of questioning that is typical of the school dialogue (from a teacher to a pupil). In this sense, he understood that the observer's question is not because she can not manage to calculate the profit but because she is interested in the processes they use to think.

Further on, the observer does not give him a specific situation to think about. Under these circumstances with other *ardinas* it was necessary to reformulate the question so as to offer a situation in which they could place themselves. Diku does not need that and organises his explanation around values (50 newspapers and then 20 newspapers) which he feels are appropriate in order to satisfy the observer's curiosity. As a good informant, who is aware of the observer's implicit interests in the questions she asks him, he also demonstrates greater ease than most *ardinas* in explaining processes explicitly. In other words, he shows himself to be more at ease in the situation of talking about the mathematical aspects of his practice than we had come to expect from interaction with the other *ardinas*.

Another distinct form in his discourse is shown in the phrase '*25 at 12.5 escudos is 312.5 escudos*' in which he uses a formulation that is not part of the other *ardinas'* usual patterns of dialogue, even with the observer. This form of discourse seems to be quite close to the usual language-game in school mathematics (form of talking in the school mathematics discourse), when relating a variable with its unit values in the way we have just seen. Diku uses the typical way other *ardinas* have for explaining this type of relation '*one is 12.5 escudos, 4 I'll earn 50 escudos*'. On the other hand, he often tries to highlight the relations between quantities (25 is half of 50) in order to justify the calculation with the corresponding value in money, thus demonstrating a kind of pedagogic attitude. But at the same time, when he was asked why he went from 16 to 20 he calls to a logic quite specific to the *ardinas'* practice: the value of 4 newspapers (50 escudos) that corresponds to a widely used coin. On the other hand, the numbers he used to explain the calculation he made to reach the profit for the 35 newspapers are also typical of this practice (the 25 in particular). So Diku seems to show that he is aware and uses the strategies typical of the *ardinas*' practice but he is also sensitive to the observer's interest in an idea of mathematics as something that has to do with relations between numbers and not only with calculation methods.

Let us now look at another episode where the same *ardina* demonstrates forms of thinking that are equally different from most *ardinas*, but in which we can identify certain characteristics that reveal not only aspects of the practice but possibly also the school framing.

Epis. 8

In 1998 two *ardinas* (Djeps and Diku) are seated outside *0 Tempo* agency waiting for the distribution. The observer is seated with them and they talk about the possibility of a profit in each newspaper being 15 escudos. She asks how much could they earn for 10 newspapers in that case. Djeps explains to her by thinking in groups of two newspapers and says '2 is 15 escudos, 4 is 30 escudos…'. He takes a longer time to say the value of 8, 50 Diku decided to intervene saying that 10 is 150 escudos. So she asks him how he thinks about it.

Diku – 15,… as 10 newspapers is,… at 12.5 escudos is 125 and it rises more 25… is 150 (*he makes a movement with his hands to show the idea of rising*).
Obs – How did you think that it rises more 25 escudos?

Diku – I did... each newspaper, 2.5 escudos for one, plus 2.5 escudos for another,.., and so (*he indicates with the fingers one by one*)
Obs – Did you add one by one?
Diku – No! I add all together... (*he makes a movement with his hands to show that is referring to the totality*)
Obs – Everything? All at once? How did you add all at once?
Diku – 10,... is 150.
Obs – And if it is 20 escudos for each newspaper? How is it for 10 newspaper?
Diku – 200 escudos.
Obs – Why?
Djeps – 'Cause 5, if it is 20 escudos each...
Diku – 5 is 100 escudos...
The distribution begins and they go to receive the newspapers.

The observer was surprised by Diku's explanation for it was completely different from what happened with other *ardinas* (who used Djeps's strategy). His way of thinking led him to link both situations, the current one and the hypothetical one, relating the current unit price with the hypothetical price that he was faced with so as to use a value known to his practice (125 escudos) to calculate what he would make faster. Curiously, when he explains in more detail how he reached 25 escudos he clarifies the difference he found between the two unit values (2.5 escudos). However, he describes the global procedure as an adding process (just like Djeps), both in words (added) and in gesture (counting one by one with his fingers). Next, with the observer's question he realises that he gave the wrong idea and tries to explain again that he thought in whole terms. In other words, he may be referring to a multiplication without, however, naming this operation adequately. He might have done so by using the word 'multiplication' (a word from the school frame) or, as Kodé did in episode 6 with Pitchiu, by using the expression 'times' (a practice word). It seems that for Diku the 2.5 escudos × 10 was immediately visualised as 25 escudos, almost as if there were a 'blind consciousness'[15] (Bloor, 1997) of what happens in the multiplication by 10, but also knowing that this change results, in fact, from the successive addition of equal numbers.

With these two episodes we intended to highlight how this *ardina* makes both 'worlds' of the situations lived with the observer – the *ardinas'* practice and the school practice – visible through small dialogues and how he lives this transition entering a new role – an informant on *ardinas'* practice. This is clearly suggested by the situation that is not common in this practice (dialoguing with an outsider of the practice who has certain features that resemble those of a teacher). We must note that this *ardina* talked (in quite good Portuguese attending to the fact that he was out of school for almost one year at that time) with the observer several times and spontaneously about school, reading and mathematics at school, things that he said he liked very much. Besides that his motives to participate in this practice were somewhat

[15] This idea of 'blind consciousness' is how Bloor, 1997, pg. 51, summarized how he read Wittgenstein expression of 'obey the rule blindly', Wittgenstein, 1992, §§: 219

different from the others. He liked coming into the 'big city' watching people and the movement of the city, to participate in a different life from what was possible in his very small village. To participate in the sale satisfied this willing in a way that was acceptable to his family. All this contributed also to his explicit preference for staying with the observer – chatting with her in the Square rather than going and selling more. We can speculate that his particular empathy with the observer drew also from seeing her as closely related with the other 'world' he wanted to participate in – a scholarly, literate and urban world. Offering himself as a good informant was one way of positioning himself in a transition experience: participating in an alien practice to him (the research practice) about a practice where he was a full participant (that allowed him to show and explain what he knew about it) and that the observer was trying to know better. But participating in this experience fulfilled his own needs and not only the researcher's (at the same time that he was aware of being helpful to her). In this sense it seems to us that the way the participant understands the relevance of the transition experience to his own life, to fulfil his own purposes or needs, can be an important element for how he will involve himself and take the best of it.

2.2. *Living in Between*

One of the main features of the *ardinas'* practice is its dynamic nature that demands adaptation to new interactions and new rules that arise from the historical transformation of the context, both from the integration of new members (a 'natural' movement within the practice) and from the changes pushed by some instances of the wider socio-economical context in which the practice is part of (that is, from the movement 'outside' the practice). Another aspect that seems to play an important role in the constitution of *ardinas'* practice is the fact that their activity is lived between an institution (trying to adjust to a market-driven modern economy) and the daily life of ordinary people (still ruled by a very traditional and community-oriented way of living together). Those two features of the practice turn out to be very important in bringing particular demands to the *ardinas* daily life, constantly confronting them with different values. The dynamic nature of the practice together with the constant living in between two socio-economical systems seems to become another instance of the transition experienced by the *ardinas* during participation in the practice.

We will now focus on the ways participation is sustained within the practice and how mathematical facts gain sense from the development of competence in the process of increasing participation from newcomer to old-timer.

2.2.1. Sustaining the participation and the practice

In the *ardinas'* practice it was particularly interesting to observe two forms of sustaining the participation in this practice. On one hand, according to the way Lave and Wenger (1991) refer to the 'sustained participation of newcomers, becoming old-timers' (pg. 57), the *ardina* can be viewed as an element of a community of practice acting in order to maintain the possibility of the existence of the community. On the other hand, when the *ardina* reformulates the institutional 'rule' in order to be able to maintain his participation in the practice, this seems to us to be very close to Wenger's (1998) call of attention to the 'non-participation developed with respect to the institution (as) an integral part of their identities of participation in their own communities of practice' (pg. 172). We shall now turn to these two aspects of sustaining the *ardinas*' practice.

In the brief introduction to the *ardinas'* practice (section 1.2.) we described what we call the 'learning curriculum' (in the sense of Lave & Wenger, 1991)[16] which existed in 1998 (with quite a lot of input from Disidori) and in 1999 (without any institutional interference). In short we may say that over these two years this curriculum had a similar form as far as the organisation of activities for newcomers was concerned (which allowed for a safe participation, both for them and the institution itself). This organisation was perfectly integrated in the vaster organisation of this practice's activities and apparently it seemed no different than other *ardinas*' normal practice. However, if we look carefully we can find certain nuances that may be relevant. For example, the fact that newcomers received newspapers from an old-timer was a sign of their participation being peripheral, albeit legitimate. In the same way, the smaller numbers of newspapers which newcomers handled at the beginning made their task less demanding both in physical terms, in keeping the newspapers in good condition, and in the level of attention that is necessary to deal with the money involved. Besides, limiting the sale of those who are learning to the Square or close to an old-timer, reduces the difficulty of these *ardinas'* tasks. Doing so, they need not pay attention to a multitude of factors involved in the practice, such as the time and place (whether it has movement or not), the pirates that might be hanging around them, the ability to prevent future situations and thus organise the change to give the client so as to have enough coins and bills for other situations. Only after some time with an old-timer do newcomers venture further and receive more newspapers. This evolution in the sense of a gradual autonomy from his partner makes the newcomer's activity more complex, requiring greater knowledge of the city and its rhythms, for example. Apart from this, it introduced greater possibility of conflicts with *piratas* or situations where the *ardina* might give the wrong change and even lose money. The consequences of many of these situations would only be felt during payment, hence the awareness of the need to frequently check the correspondence between the newspapers sold and the money in their pockets. In general, stories

---

[16] 'A learning curriculum consists of situated opportunities [...] for the improvisational development of new practice. A learning curriculum is a field of learning resources in everyday practice viewed from the perspective of learners' (pg. 97).

'Production activity-segments must be learned in different sequences than those in which a production process commonly unfolds, if peripheral, less intense, less complex, less vital tasks are learned before more central aspects of practice' (pg. 96).

about this kind of issue were told right at the start, but the newcomer would only gain the habit of actually checking much later in their learning process.

In 1998, deciding which old-timer would be responsible for a newcomer was Disidori's decision but this was well accepted by all the *ardinas* ('I like to help people' they would say) and it was even a way of acknowledging their status inside the community. In 1999 the social organisation of the *ardinas*' practice had to adapt to the birth of another newspaper (*O Espaço*). In the initial description we spoke about the main changes that occurred but it is now important to point out certain aspects, particularly in relation to the newcomers' integration. While attempting to keep up a learning system for the newcomers (a part of the history of the practice) by themselves, the old-timers would adapt it to the current situation and show how they felt regarding the newcomers' integration. The entry of newcomers increased the number of *ardinas* selling, which could mean that each one would earn less money. When the old-timers took on the role of helping the newcomers, for a time they placed themselves in a relatively powerful position as gatekeepers of the community. This power helped them to increase their gains by establishing a price for their 'help' and simultaneously to control a situation that had the potential to threaten their own place. The internal acceptance of this situation also shows how the importance of a privileged relation between an old-timer and some newcomers is acknowledged (both from old-timers and newcomers), at least in the beginning. The newcomers face a new and sometimes frightening situation (in the relation with the *piratas*, in handling a lot of money and in the responsibility this involves) and have the comfort of 'friend's' protection and help. Actually they usually referred to a need for protection rather than for learning. As we noted before, the learning situation was not limited to the privileged relation between an old-timer and a newcomer; mutual support between peers (other newcomers or else slightly older with little experience) was important also in sustaining the newcomers' participation. This adjustment in the learning system not only reveals the dynamic character of the practice but also the agency of *ardinas* as a community (before the institution) in which certain elements have the ability to take on a power[17].

Sustaining the participation has another important aspect which emerges from confrontation between the *ardinas'* need to maintain the possibility of making money with some regularity every week and the rules defined by the institutions with which they interact in their practice. For example, some of those who intended to sell in certain spaces (public services, banks,...) took the care to dress and behave in a particular manner. However, they all realised that somehow their participation had to take place within the social organisation of the practice, particularly regarding the norms of the agencies whose newspapers they sold. When we tried to identify the fundamental lines of this organisation from the *ardinas'* point of view, we came face to face with the strong presence of the idea of 'rule', both in their discourse and in their behaviour. We realised just how strong these rules were and how they sometimes sustained the *ardinas'* participation. That is, when these rules were made

[17] 'Understanding power as collective agreement conferring authority' (Wenger, 1998, pg. 15).

visible and we understood their inclusion in the *ardinas'* activity (considered to be rules in the sense that they help also to constitute the practice) and how they interacted with the organisation imposed by the institution. Some of these rules were very explicit and institutional (in the case of *O Tempo* in 1999, or the obligation to pay for the sold newspapers), others less, arising at times from interactions between *ardinas*. Although their most natural attitude was to accept and respect the enforced norms and hierarchies, sometimes they needed to 're-read' these norms or adjust their procedures so that they would appear to follow them. It was a way of the *ardinas* themselves managing some of their needs in the strictness that was imposed to the *ardinas*' group. For example, how they modified (in action) the rule of *O Tempo* '*not to sell O Espaço when selling O Tempo*' transforming it into '*not to make visible O Espaço newspapers when going to get O Tempo*' made it possible for them to sell both newspapers at the same time, increasing their ability to make money but also to sell either of them faster. Therefore, even in the case of those rules that were defined by the Institution, the *ardinas* would adjust them, on the one hand, for sustaining their participation in this activity but, on the other, for sustaining the practice that should continue for them to satisfy their main goal. At the time, without a set of *ardinas* who would constitute a community of practice (guaranteeing the newcomers' learning) it was in fact not possible to sell newspapers in Praia. So it became a mutual need (for the *ardinas* and for the newspaper's administration) to maintain this practice and through it to sustain the participation of *ardinas* in the practice. It was possible to notice that some *ardinas* were selling both newspapers at the same time. The ones who bought *0 Tempo* (and not take) maintained *O Espaço* visibly when they went to *0 Tempo* board to receive their newspapers. However nobody there argued about that situation with them. Things were left in the realm of the unsaid, of what is not convenient to clarify completely since it involves risks for the institution itself. Situations of disruption did no good to either of the sides so many situations did not lead to a point of total rupture, whether by the newspaper (Disidori, Anriki, newspaper board, Manu) or by the *ardinas*.

As Wenger (1998) suggests, 'such a concept of practice includes both the explicit and the tacit. It includes what is said and what is left unsaid' (pg. 47). It seems as if in periods of main changes occurring within a practice the need for sustaining both the participation (and thus the community of practice) and the practice emerged more strongly than in moments where the changes evolved more naturally from the internal evolution of the system (as it was in the case of the daily integration of newcomers).

2.2.2. 20 is not the same as 100 – 80

As we mentioned before, in March 1999 there were, at the same time, two different forms of *ardinas'* access to the selling process of *O Tempo* ('taking' and 'buying') but all the *ardinas* took newspapers in *O Espaço*. Some *ardinas* experienced both and they lived those two systems when they sold both newspapers. Within the taking system, *ardinas* were allowed to receive a certain amount of newspapers, to give back to the agency the newspapers they were not able to sell and to pay for only the sold newspapers at the end of the selling process. However, some *ardinas* were obliged to buy the newspapers they received if they wanted to participate in the selling (as was the case of some old-timers from Praia). From the start all *ardinas* were on the regime of taking; however, when one of them did not pay in time (thus showing they were not serious), *O Tempo* changed the system for those particular boys. Usually, two or three weeks after, these *ardinas* would come back to the agency asking to be allowed to sell again. If they were accepted, usually they were integrated in the new system (buying). In either system, the money earned from each newspaper was the same (20 escudos). But when the *ardinas* took they had to go back to the newspaper agency to return the unsold newspapers and pay for the ones they sold, but if they bought they should pay 80 escudos for each of them as soon as they received them.

Let us now see how the *ardinas* viewed this situation. We begin presenting three episodes, which will be discussed jointly, despite having occurred at three different moments. They involve four *ardinas* who had been serious regarding payment procedures with *O Tempo* up to that point. They sold newspapers on a taking regime and were attending the school in the 6th grade. Ntóni was the youngest in the sale, Djoka and Lulu were also young but had more experience than Ntóni, and Manitu had already begun to sell in 1998. We shall also comment on the opinions of certain old-timers about the same issue. The dialogues happened with the observer and dealt with the question of which system they considered the best. Therefore, the problem was not expressed with the intention of assessing whether these *ardinas* thought the profit of 20 escudos in both systems was identical. Because the profit was the same in both systems (20 escudos) what we intended was to understand just how relevant this mathematical fact (20 = 100 – 80) was considered to use as a base for their choice between the two systems.

Epis. 9

Ntóni and Djoka are discussing which newspaper is better to sell and what is the best access way to sell (buying or taking). The observer is the one who provokes this subject. They disagree about it, Ntóni saying that buying is best and justifying like this:

Ntóni – When we take newspaper we can't take off our money…
Obs – Don't you, the profit? When you take, how much is the profit?
Ntóni – 20 each.
Obs – It's 20 each newspaper? And when you buy them?
Ntóni – Pay 80 and sell 100.
Obs – 100, and the profit? (*Djoka says that is 20*) It's also 20? Is it the same thing? (*Ntóni stands still but Djoka nods yes*) But,… why is it best to buy?
Djoka – (*makes a movement with the hand towards his body*) Taking.
Obs – Is it best to take?…
Djoka – If you don't have money…

Epis. 10

The dialogue is between the observer and Lulu, in the Square, after finishing the payment of *O Espaço* he sold. They are both seated on a bench, together with Manu and Ntóni that observe them without speaking.

Obs – (*to Lulu*) Did you take or buy them?
Lulu – Take.
Obs – Take, hum.., and what thing is best, to take or to buy?
Lulu-Buy.
Obs – Buy is best? Why?
Lulu – 'Cause is yours... (*he points to himself*)
Obs – But do you earn the same? (*he nods no*) No?! Do you earn more? (*he nods yes*) How much? When you take how much is it?
Lulu – When I take earn 20 in 100, to buy is... 30...?

He hesitates a little and looks in the direction of Ntóni and Manu as if he is asking for some sign of agreement. Ntóni responds to his demand saying that it is the same.

These two episodes allow us to reflect upon two aspects: (i) the choice made by two of the *ardinas* and the way they begin to justify this choice; (ii) the apparent difficulty in clearly expressing the real profit in each regime. Ntóni and Lulu perceive the 'buying' system as having more advantages and express essentially sociocultural (and not mathematical) reasons for this. For example, their first justification seems to highlight a concern for a sense of possession ('*you can't take our money*' and '*it's ours*'). On the other hand, both show doubts when clarifying the profit in each of the regimes. Ntóni is able to explain the profit correctly when buying by using a formulation that could indicate an ability to understand that it is, indeed, 20 escudos. Yet quite unexpectedly he reveals difficulty in realising that the profit made in both regimes is one and the same. Lulu, on the other hand, does not seem to know how much is earned in the buying regime. We can interpret these hesitations in two ways. Since neither of them has experienced the buying regime they do not yet know the complexity involved 'from the inside'. In fact, for the *ardina* the taking regime is less risky, for it guarantees an effective profit of 20 escudos for each sold newspaper, unlike the other regime where this profit is smaller if the newspapers are not all sold. This also seems to reveal that these aspects are not objects of conversation between the *ardinas*, they must live the experience in order to actually 'learn' which variables are involved and the importance each one must have for making decisions. But it is also possible to think that the fact that the sense of possession is more evident in the buying system may be relevant and lead to a situation where the 'objectivity' of $20 = 100 - 80$ (a mathematical fact) is meaningless for them in the light of social references.

The third *ardina* involved in these dialogues (Djoka) raises another element ('*if you don't have money*') in his justification, which we also found in other *ardinas* (who were more experienced than Ntóni and Lulu). This seems to support our previous interpretation. It is the case in the following episode.

Epis. 11

Later in the same day, the observer has a long conversation with Manitu as he sits near her in order to rest a little and showing interest in talking with her. They talk about several aspects of his *ardina* life but also about his family and the school. The observer felt that it could be an opportunity to talk with him about the two systems (taking and buying).

Obs – What thing is best? Take or buy newspaper?
Manitu – Take, for instance,... if one time you don't have money to buy you stay as,... you can't sell (*he says this last idea in a slow manner as if he was showing sadness*).

Both Manitu and Djoka justify their decision for 'taking' as the best regime for those who do not have money, relating it to the access to selling. They seem to suggest that this situation is seen as a sign of inferiority and dependence. In this case they have to take advantage of the organisation that is institutionally imposed so as to guarantee their participation and, therefore, satisfy the need that led them to participate in this practice in the first place. On the contrary, in the buying regime there is a possession of something from the start. When buying the newspapers at the agency, the *ardina* has a reasonable amount of money from the start that he uses to exchange for something (the newspapers) and from that moment on all the money he makes will be his. In turn, the *ardina* who takes newspapers receives them without giving anything in exchange. At the end of the sale of two *ardinas* in different regimes, the way each of them faces the others (and it is seen by the others) is really different. For the *ardina* who takes, the act of paying is a moment where he delivers a great amount of money, leaving a very small part for himself. On the contrary, the ultimate image of the buying *ardina* is of possessing a great deal of money, all he got during the sale.

Curiously, there never occurred to the *ardinas* an opposite (but possible) interpretation. In fact, when the institution delivers newspapers in exchange for nothing, they do not see this as an attitude of trust towards the *ardina*, as an acknowledgement that the person is trustworthy. *O Tempo* introduced this regime of buying in two types of situations. At the beginning it was used to distinguish between the Praia *ardinas* and those from S. Martinho (as a punishment for these for having a privileged relationship with Disidori with whom the administration was on bad terms). Later it was adopted as a way of dealing with those who did not show up at the end of the sale to pay. The latter is an attempt to reduce the risks surrounding the delivery of newspapers to youths coming from families with economic difficulties without having to control them more directly (in 1998, done by Disidori). This attitude of *O Tempo* was interpreted by the old-timers as a sign of less solidarity and of competition, both between groups (S. Martinho and Praia) and between newspapers. On the other hand, they realised (some of the old-timers said it clearly to the observer) that it was a way the newspaper could get the same service for a smaller cost for it did not need to pay an employee to perform Disidori's role. The *ardinas* felt that it is better to take than to buy, since, for example, '*when you don't sell everything the agency recovers whatever is left*' or (regarding the buying regime) '*you've to be careful and not ask for many 'cause you won't sell them*'. Besides this, they clearly say that '*this year O Tempo doesn't trust the* ardinas' and present this as one of the justifications for not complying with their obligation to pay. In 1999, the *ardinas'* daily practice runs amidst two systems whose organisations reveal the presence of values which, in one case, have a more traditional character and, in the other, are close to those of a mercantile economy. The way the *ardinas* behave with each of

---

[18] For instance, when *0 Espaço* newspapers are not ready in time *ardinas* like to help the workers inside the board even without payment for that work, but they do not do the same with *O Tempo* (in 1999). Similarly the *ardinas* behavior towards *O Espaço* and *O Tempo* related to the payment are different. Old-timers sooner or later stop being serious to *O Tempo* regarding the payment of newspapers sold but never to *O Espaço*.

the newspapers[18] shows, in several moments, how they perceive this dissimilarity, how they interact with it and how they organize themselves in order to satisfy the needs underlying their integration in this practice.

At several moments of the two periods of data collection, when asked to characterise what they thought was '*a good ardina*' all of them started by saying that '*he should follow the rules*' or '*be honest*' referring most of the times to the payment rules. After that, they accepted that he should be able to sell quickly, know the best places, address the client correctly and finally give change. Besides what was actually said by the *ardinas* when questioned, most of them also had certain behaviours that revealed a great concern in being correct about payment procedures. The act of paying the sold newspapers structured their way of living during the whole sale time (as was described in 1.2). On one hand, it was during these moments that more collective situations were experienced, where behaviours came out and stories of the day were shared. Although largely lived on the streets, these were the most structured moments of their interactions with Disidori and Manu, and the procedures were more explicit. In fact, these were the only moments when cultural artifacts of the practice were used in a consistent way – the calculator and written records on which *ardinas* were selling on that day and how many newspapers they had taken. Besides, this was when large sums of money were visible (the money is in fact the first appeal that leads *ardinas* to this practice). These circumstances (artifacts and conditions of the activity) helped confer a certain solemnity and respectability to the moment, that is, they helped define the status of this phase in relation to the others (distribution and sale) and were genuine structuring resources (Lave, 1988)[19] of their selling activity. The newcomers, in an increasing participation (where the peripherality was legitimate so it was positive), also have access to this 'world' of payment, learning to recognise its importance. In 1999, the visibility of payment procedures at *O Tempo* as a component of a community's practice, was more diluted. Requiring a new trip to the agency after satisfying the first goal of the *ardinas'* activity (making money from selling newspapers), as well as the fact that there is one group which does not have to do this (mostly the old-timers) also hinders the wish to perform well. Starting with what we could refer to as a mathematical fact that the *ardinas* knew (20 as a result of the subtraction 100 – 80), it is interesting to see that they actually even refer to a lower profit in one of the cases, based on essentially socio-cultural justifications. They become fully competent in the practice when their participation in it is a 'full participation' (Lave & Wenger, 1991, pg. 37). In that process of increasing participation they are, for instance, using the resources that structure their activity, experiencing various situations which arise from the practice's organisation, therefore understanding the diversity of relations involved in it. Only then do they compare the two systems of access to the practice with more 'objective' references, which raise the fact that the profit might not be exactly the same in the two situations. In other words, the competency of being an *ardina* is the

[19] Santos and Matos (1998) discussed and analyzed the difference between structuring resource and cultural artifact and their role in school mathematics learning.

starting point for building bases for the knowledge (or re-contextualisation) of facts, in this case mathematical facts.

It seems that the apparent less visibility of control on the part of the agency (*O Tempo*) made stronger the need for cohesion among the *ardinas'* group. This led them, for instance, to re-write the institutional rules in order to avoid colliding with the institution but making it possible to keep a certain sense of belonging to a community. Besides that, the *ardinas'* greater autonomy (also apparent in not being pointed out by the authority – any old-timer could help the newcomer) made clear the need they felt for a certain learning/protection system that helps the newcomer but keeps the control within the gatekeepers of the community (the old-timers, protecting the individual *ardina* but also the community, sustaining both the participation and the practice). This situation came in a period of great instability within the practice as a result of the co-existence of two different sets of rules (in the two agencies) and, in one of the agencies, two different systems (buying and taking), together with a more visible effort from this agency to move forward to the market-driven economy emerging in the country. That is, the *ardinas* were pushed to live frequent transition experiences in a practice that is no longer ruled only by the agencies or by one system, but where they all together constructed their own rules, the context of the practice, as if they were living their daily activity in a kind of transitional space. It was possible to observe different ways of dealing with those transition experiences particularly between the newcomers and the old-timers and we could realise how they interrelate, for instance, with their perceptions of facts that apparently are objective (as was the case of the amount of profit). On the other hand, the contribution that the participants recognise in their participation to their own life projects seems to constitute an element that plays a role in the way they engage on the transition experiences and how they bring in the resources available not only in the practice where they are situated but also from others practices where they had participated.

3. WHAT DID WE LEARN?

A powerful idea that we would like to bring to the fore again is the understanding of 'learning as an integral part of generative social practice in the lived-in world' (Lave & Wenger, 1991, p. 135). One should recognise that this means that participation in social practices does not merely influence otherwise autonomous psychological processes. Under this perspective, learning means changes in the ways that a person participates in social practices. So, it seems to us that if we propose ourselves to understand how mathematics learning connects to social practices where mathematics' use is present the social practice is the 'primary, generative phenomenon and learning is one of its characteristics' (Lave & Wenger, 1991, p. 34).

Our main concern in the whole research project was driven by the need to better understand the relation between social practice and (mathematics) learning. With this main goal we decided to observe, describe and analyse a social practice where we (and the participants) identify the use of mathematics (even with different meanings). In this work we need to situate that social practice within the social world where usually it takes place but also to have some elements that allow us to analyse how the participants evolve through the historical transformation of the practice. Through the work already done in the main project we confronted ourselves with some issues that emerged from the analysis of particular moments of the *ardinas'* daily practice – those we understood as experiences of transition that took place, some within the historical transformation of the context of that practice and others because of the research situation. Focusing the analysis on those moments appears to be fruitful in bringing to the front some aspects such as: (i) the relation between the boys' participation in the *ardinas'* practice, their learning and use of mathematics in that practice; (ii) the connection between the mathematics of the practice and the school mathematics and its presence (or absence) within the *ardinas'* practice.

From the very beginning we stressed the analytical tools that enabled us to go deeper in our research (some of them may not be very visible in this chapter although they played a role in the preparation of some discussions). For instance, in order to describe and analyse the presence, meaning and importance of rules in the *ardinas'* practice we were relying on Wittgenstein (1992/1953) perspective about rule-following and about meaning as grounded in collective use, as well as on Goffman's analytical framework of rules (particularly on what he called regulative and constitutive rules). The study of Goffman' s social theory went in parallel with our reflection on research questions and data. And this led us to find as relevant Goffman's (1991/1974) rationale on frame analysis. This second aspect of his approach to the study of interaction order played an important role in our use of Restivo's idea of thinking as internal conversation (different from external conversation) as one way of making sense of differences identified in the *ardinas*' discourse in some particular moments. Our interest in the approaches of Restivo (1998) to thinking and Julien (1997) to competence came from our effort to understand how *ardinas'* mathematics knowledge emerged from their participation in the practice.

Along with our concern with those issues about mathematics learning, several methodological issues turned out to become relevant and part of our reflection on our research practice. The type of research questions together with the theoretical and analytical approach we took, suggests that we present what we have learned from our own participation in the study mainly around two aspects: (i) what it concerns with regard to the perspective of learning as participation in a social practice; (ii) what relates to the methodological aspects of the research.

3.1. About Learning as Participation in a Social Practice

With the analysis reported we intended to give evidence of changes in the forms of participation of the *ardinas* in a practice and therefore to show instances of learning. We will now summarise our findings mainly in two aspects that appear to be the most fruitful ones to contribute for expanding our understanding of mathematics learning as (legitimate peripheral) participation in a social practice: (i) what emerges when we look at *ardinas* practice as a social practice; (ii) what comes when we look at *ardinas* as individual participants in that practice.

3.1.1. *Ardinas'* practice as a social practice

The analysis showed evidence of the **dynamic** character of *ardinas'* practice that came from changes that took place both around and within the practice, and that plays a role in the way participants participate in it. On one hand, the arena as a preexistent entity independent of the participants has its own evolution according to forces and relations that emerge from its nature (an economical institution) and position within the society as a whole. It was the case, for instance, of the existence (in 1999) of two newspapers instead of only one (in 1998), or the change of the person who used to control the *ardinas'* work. On the other hand, as the *ardinas* only had an informal link to the agencies and most of them didn't integrate their participation in this practice within a professional perspective (it was socially regarded only as a work for young boys and not as a profession for adults) they tended to participate in it for short periods of time in their lives. So, within the practice there were always some newcomers arriving and some *ardinas* at different stages of competence or even changing roles. In this way we can say that their community of practice was not a rigid or closed one although a certain control of the access to it was carried out both by the two newspapers' institution and the *ardinas'* community.

Another aspect that seems to be visible when we look carefully to the *ardinas'* practice is its **non-self-contained** character. That is, what was happening there, how participants behaved within it, how they read what was important or not, or what they used as resources for taking decisions was shaped not only by what was particular to that practice but also by the connections to adjacent practices. For instance, the presence of values was clear, sometimes conflicting but co-existent, coming both from the economical character of the practice and from the more traditional humanistic aspect of the culture of Cabo Verde that shapes the relationships between adults and children. Both those different values had a role in the constitution of the practice, shaping the participation and the *ardinas'* reading of what was important or not within it. As a social practice (that took place in a mainly socio-interactional and open environment) the social references were more visible than the mathematical ones. Therefore when *ardinas* needed there to decide about what was more appropriate, they relied more spontaneously on the former than the latter ones.

Another interesting aspect of this social practice that seems to play a role in the *ardinas'* learning is the **history** of the practice; how it passes to them, how they are aware of and appropriate it; how they re-write it and, in a sense, how it organises also their learning curriculum.

From these three aspects of the practice we can conclude that when the *ardinas* act in order to sustain participation in the social practice they are living moments of transitions (from newcomers to old-timers, from one role to the other, between different rules and values) within a certain historical recursive (but not equal) reproduction. These seem to us to be central issues that helped them grow as competent learners and not only as competent *ardinas*.

3.1.2. *Ardinas* as participants (within a social practice)

An important aspect that emerged from the analysis reported was the pivotal role played by the *ardinas*' effort **to sustain both their participation and the practice**. Even the main (and initial) goal of their participation being the need to help the family by earning some money, some other particular motives (to feel the movement of the big city, to have money for gambling) shaped their approach to the practice. On one side, they wanted to maintain their acceptance by both the agencies but also by the community, so they needed to learn how to go on with both of them. This was the case, for instance, in the input they had (in their action as a community of practice) in the adjustment of the rules, but also how they individually adjusted their own behaviour to the possibilities they 'read' as acceptable in each of the agencies. On the other hand, in their movement from newcomer to old-timer, some of them changed their roles, for example 'teaching' one newcomer or being in control of a group of *ardinas*, or even being involved in activities within the agency. In this transition, some of them acted more strongly as the gatekeepers of the *ardinas* community and others more as 'representative' of the agency. However, none of them put totally in risk their participation both individually and as a group, so in a sense they acted also in a way that ensured (to the agencies and the customers) the utility of the selling process of newspapers being done by a group of *ardinas*. In a way there was among the *ardinas* a sense of having some power both within their community of practice and in relation to the social role played by the newspaper in the society.

As we saw the *ardinas* as individuals have different motives for participating in the practice, but they also show different **ways of dealing with the resources** available within the practice. This is the case not only in the **ways they appropriate the resources** (as support elements for their calculations) but also how they connect them with other resources available (for example, their knowledge from school mathematics or from other newspapers' selling activity). The analysis showed evidence that it was only when the *ardinas* get a certain competence in the whole practice – according to Lave and Wenger (1991) when they achieve full participation in the practice – that they are able to use the resources in a more effective way for 'reading' the situations. This was the case, for instance, in comparing the two systems of access to the practice with more objective references that bring into play the fact that the income can be different in the two systems.

Additionally, the competence of being an *ardina* (in what it involves to be familiar with the usual calculations involved but also to know and accept the values of the practice as their own) makes a difference in **what they elect as a problem and how they respond to it**. For instance, when we were analysing their internal conversation it was possible to recognise the existence (for the *ardinas*) of different problems in situations they usually referred as the same 'to check the money and newspapers'. So the meaning (in use) of this expression varied according to their competence. The strategies they used to deal with it were not always the same, it could be counting (money as newspapers) or calculating (the money they should have for paying back to the agency). It was the case for the newcomers, who carried few newspapers with them, that they did not buy anything before paying back the newspapers but they also knew that sometimes they lost money or newspapers. So, their need for checking usually came to their mind as a suspicion that someone took some newspapers from them or they could have given some change incorrectly to a customer. This is a newcomer problem and they use more a strategy of counting the total amount of money and newspapers they carry all together. On the other hand the old-timers sometimes do buy something before paying back to the agency and they are confident that they did not lose money or newspapers, so what they try to figure out when they are checking the money and the newspapers is to see if they have enough money to pay back the newspapers already sold. In this case they do not count or think of the total amount of money involved in the selling but only of the money they should have in order to pay for the newspapers. The focus of their problems is on being able to follow the rules of being a good *ardina* (paying back correctly). In this situation the calculations they do are much more complex, and show a shared repertoire of routines using what we called the same support elements for calculations.

The basis for knowledge (or for the re-contextualization) of (mathematical) facts seems to draw from the competence of being a participant (*ardina*). That is, being able to act in order to sustain the individual as well as the group participation and at the same time to sustain the practice itself, to elect what are their problems within the situations and according to their ability to use the resources from that particular practice as well as from other practices where they participate, to deal with those problems in an appropriate way (in what this means within the practice and for themselves). This seems to be a relevant issue if one wants to understand the ways students learn within their participation in school mathematics practice in the classroom. Learning to understand school mathematics practice and the role of that practice in the students' life projects from the point of view of the learners is the starting point for the analysis of how students learn school mathematics.

3.2. ON THE RESEARCH PROCESS

The methodological difficulties of gaining access to children's meanings are visible in the study reported. The fact that the research is studying a phenomena which was almost totally strange to us in most of its aspects, led us to realise that we had to go through a process which should involve, to a certain extent, our participation in the (*ardinas*) practice with the explicit (for us and for them) goal of learning it but not in order to be a full member of that community of practice. This starting point (more in terms of knowing that there are more things that we don't know than that we know) opened that community of practice to us but also gave us consciousness that methodological issues were central in this research.

First, there was the recognition that we were outsiders in a practice which was not familiar and that we couldn't understand in full. There were several issues contributing to this: (i) the difference in languaging among the *ardinas* and when they talked to us; (ii) the fact that we were white people in a black community; (iii) the fact that there was difficulty at the beginning of thinking in terms of a radically different practice (selling newspapers against teaching mathematics), particularly with regard to the role of discourse within both practices. An important fact was that the researcher collecting data in the field was an adult woman and she was entering into a practice developed mostly by young boys and very few men.

Because there was a strange person in the group the need to create an emphatic relation with the group became relevant and explicit. But at the same time there was outside interference in the practice that was taking place. Because one really wants to enter in that practice, empathy is being created but this leads to a change in the practice – or at least in the way the practice is practised when the observer is present.

There is evidence that the young boys (observed within their practice as *ardinas*) work in different frames according to their motives, needs and values at the moment. By interacting with the observer as informants they were often brought into a frame different from the one they used in their ordinary *ardinas*' practice. In particular, when they tried to explain to the observer their processes in dealing with the calculations within the practice they tended to put themselves in a frame near that of schooling. This could have happened because it was the frame they imagined she could better understand (she was a woman as usually their teachers were, speaking more fluently in Portuguese, which is the school language) or because their practice was very little argumentative about those issues both within the community and with the authority (the people from the agencies who control their activity). In this situation they were acting and talking as informants in order to help the observer (an outsider) to make sense of what could be understood as the mathematical knowledge developed within that practice. Doing so some *ardinas* showed that they were dealing with two different definitions of the same problem but others acted as if there were two different problems. In both cases they were living a non-ordinary experience within their usual practice – talking about mathematical facts. They showed difficulties dealing with the need for re-framing their practice in terms of the frame of schooling. This was clear in the discourse of the *ardinas* and shows the difficulties that one can expect when trying to analyse the ways children do school mathematics and to interpret their utterances. In these situations it can be more useful to look at those difficulties not as the child's inability towards the mathematics itself but emerging from the transition situation experienced by a particular child or a particular group.

At the same time, collecting data in the street is something completely different from doing so within the classroom. In fact, in the street both the physical and the social organisation of the practice (as that from the *ardinas*) is lived in a public and non-structured (for that practice) space with different frames constantly interfering at the same time. This aspect can present problems particularly for a teacher mainly used to the school context, but also constitutes an interesting opportunity to understand the transitions people live within their practices.

A final important issue relates to former – although recent – forms of colonisation of the country of the researchers over the country of the participants; this is something that people of both countries seem to be trying to solve through relations of work and cooperation. However it is an issue that would deserve scrutiny if one wants to see the implications of this kind of research.

REFERENCES

Bloor, D. (1997). *Wittgenstein, rules and institutions*. London: Routledge.

Goffman, E. (1991/1974). *Les cadres de l'experiénce*. Paris: Les Éditions de Minuit.

Julien, J.S. (1997). Explaining learning: the research trajectory of situated cognition and the implications of connectionism. In D. Kirshner, & J.A. Whitson (Ed.), *Situated cognition: social, semiotic and psychological perspectives* (pp. 261–279). New Jersey: Lawrence Erlbaum Associates.

Lave, J. (1988). *Cognition in practice.* Cambridge, USA: Cambridge University Press.

Lave, J., & Wenger, E. (1991). *Situated learning: legitimate peripheral participation*. Cambridge, USA: Cambridge University Press.

Restivo, S. (1998). *Mathematics; mind and society: an anarchist theory of inquiry and education.* Personal Communication at 1st Conference on Mathematics, Education and Society, Nottingham 98.

Santos, M. & Matos, J.F. (1998). School mathematics learning: participation through appropriation of mathematical artefacts. In A. Watson (Ed.), *Situated cognition and the learning of mathematics* (pp. 104–125). Oxford: Centre for Mathematics Education Research.

Wenger, E. (1998). *Communities of practice: learning, meaning and identity*. Cambridge, USA: Cambridge University Press.

Wittgenstein, L. (1992/1953). *Philosophical investigations* (3rd ed.). Oxford: Basil Blackwell.

CHAPTER 5

EXPLORING WAYS PARENTS PARTICIPATE IN THEIR CHILDREN'S SCHOOL MATHEMATICAL LEARNING: CASES STUDIES IN MULTIETHNIC PRIMARY SCHOOLS

GUIDA DE ABREU, TONY CLINE AND TATHEER SHAMSI
Department of Psychology, University of Luton

1. INTRODUCTION

In industrial and post-industrial societies the process of starting school is, arguably, one of the most significant transitions children make in their development after birth. This is the moment when for the working day they leave the cocoon of the family and are placed in the care of the school, an agency set up by society to manage the next phase of their preparation for their role as adult citizens. Symbolically parents pass responsibility over to teachers, and these professionals become the custodians of key elements of the knowledge and understanding that society requires the children to develop. But parents still retain key responsibilities and continue their emotional ties to their offspring. So all too often the alliance between parents and teachers over children's learning is an uneasy one. Even in areas where teachers might expect to have unchallenged sway, such as the core subjects of the school curriculum, the influence of the home and its local community may remain powerful and may prove to be in competition with the influence of the school. Children's representations of aspects of the curriculum will be affected by what their parents and their teachers say and do and by the degree of consistency and harmony between these key players in their lives and between their actions and their words. In this chapter we aim to explore the specific role of parents in children's experience of one aspect of this transition – the transition between mathematical practices at home and mathematical practices at school.

The chapter focuses on a recent research project on mathematics learning in multiethnic primary schools in England[1]. We investigated the experiences of high

[1] This research project entitled 'Mathematics learning in multiethnic primary schools' was supported by an ESRC – Economic & Social Research Council / UK, Grant (R000222381). This chapter was written while the first author was a visiting scholar at the Department of Social and Political Sciences at the University of Cambridge, UK (February–July 1999). We are grateful to the parents, the children, the teachers and other staff in the schools for their collaboration in the project, to Maria MacIntyre for her help in the transcribing of interviews, and to the Project Advisory Group (Helen Abji, Imtiaz Chaudhry, Gerard Duveen, Zafar Khan and Terry Redmayne) for their advice at various stages of the investigation.

G. de Abreu, A.J. Bishop, and N.C. Presmeg (eds.), Transitions Between Contexts of Mathematical Practices, 123–147.

achievers and low achievers from two ethnic groups, White-British and Pakistani-British, in an attempt to understand the dynamic factors that influence individual differences among children from the same ethnic group. On the assumption that children's parents and teachers are likely to play key roles in these social dynamics both were interviewed. In this chapter we analyse the dynamics related to parents. Our interest in investigating the influence of parents in collaboration together grew out of earlier research we had carried out separately with children from home backgrounds that were culturally different from that assumed in their schools. Our starting points included Cline's (1993, 1998) work on the school situation of children with learning difficulties who were learning English as an additional language and Abreu's (1995a, 1995b) research with children from homes where parents practised a type of mathematics that was different from the mathematics of the school. In each case our earlier work had highlighted situations in which the transition between home and school posed particular challenges for the young learner.

In this study interactional effects became salient when the perspectives of both the parent and the child were considered. It appeared that parents both influenced and were influenced by their children's participation in school mathematics. So, in reporting our research we will try to make visible the representations of both parents and children and illustrate how these interact to shape patterns of transition between home and school mathematics. The children who were most successful in mathematics at school tended to have parents who were confident in helping them to manage that transition.

2. BRIEF OUTLINE OF THE THEORETICAL APPROACH

From the outset this programme of research followed a sociocultural approach informed by Vygotsky's (1978) theory, situated cognition anthropological theory (Lave, 1988), social representations and social identity theory (Duveen & Lloyd, 1990). In this approach the mathematics that is practised in distinct social practices is based on forms of knowledge that have been historically produced, transmitted and transformed. Thus its representation has a double character. It is the representation of something (and therefore a cultural tool) and of someone (and therefore seen as belonging to specific social groups) (Duveen & Lloyd, 1990; Abreu 1993; Abreu, 1995a). In adopting this perspective we aimed to explore issues related to the mastering of specific cultural tools that are required in specific contexts of practice. But in addition we aimed to articulate the impact of valorisation and identification processes on the transmission and learning of knowledge.

The notion of valorisation we follow in our studies was introduced by Abreu (1993) to explain the relationship between home and school mathematics in a farming community in rural Brazil. She argued that understanding of how particular social groups learn, use and transmit knowledge requires consideration of the link between knowledge and values. In her view it is the association of mathematical practices with particular social groups that provides the framework for understanding the value groups attach to their own mathematical practices. Social groups are located in particular social orders and do not function independently from the wider society. For example, the yardstick to evaluate and value farming mathematics against school mathematics in Brazil was provided by the access these forms of knowledge gave to individuals in the wider Brazilian society. This meant that social representations of mathematics give the individual understanding of the tools and of the codes to compare and categorise co-existing forms of mathematical knowledge.

The notion of identification aims to explain the emergence of differences between individuals exposed to similar practices. What processes lead individuals who share social representations to participate in quite different ways in the practices that are represented? This is a question about the interface between knowledge as represented in the social sphere and its re-construction in the process of individual development. Duveen and Lloyd (1990) have argued that the re-construction of social representations at a psychological level involves active elaboration of social identities. To explain this agency of the individual they distinguished as components of identity between the knowledge of social semiotic codes (tools and their social value) and the position he or she takes towards them. Positions are by definition evaluative. Individuals explain their positions in terms of how they feel towards something, how much they like it, the importance they judge that it may have in their lives, etc. However, how individuals adopt a particular personal positioning, which then becomes a part of their identity, is an area that needs further investigation.

Next we outline why we see this approach as important in the context of research on how parents support their children's transitions between home and school. The double character of social representations needs to be taken into account when investigating ways in which parents can influence how their child manages the transition between home and school mathematics. Firstly, this requires understanding of their use and/or their preference for specific mathematical tools and the way in which these can influence their children's transition. For instance, do parents tend to use mental or written tools for arithmetical calculations? When they use these tools do they emphasise rote use and memorisation or flexible use and understanding? It is often assumed in the literature that when home practices promote forms of thinking that are similar to the ones required at school, this will smooth children's transition to school and contribute to higher achievement there (Bernstein, 1973; Gallimore & Goldenberg, 1993; Heath, 1983; Tizard & Hughes, 1984). This type of explanation has been supported in various studies of home-school literacies. Recently Gallimore and Goldenberg (1993) reviewed a series of studies in which they had investigated literacy practices in Latino families in which American-born children were being brought up by foreign-born parents. Their observational and interview data convinced them that although members of the family usually valued and were prepared to help the child to engage in literacy activities what they actually did with the best of intentions was not optimal as a support for school progress. They found that a key factor in the way the activities were framed was the parents' representations, or what the authors referred to as 'scripts'. Even when the researchers tried to influence the home activities positively by creating external demands they found that 'as soon as the parents construe an activity as the "teaching of literacy", their prevailing conception of literary development is activated, driving the interaction and determining the script-in-use' (pp. 328–329). For instance, they found that parents overemphasised correct answers (e.g. reading a word accurately) to the detriment of reading for meaning.

Secondly, it is necessary to take into account the valorisation of the social practices. Thus, parents can influence the transition through their selection of the information that they intentionally expose to or hide from their children. In this case the salient aspect of the representation on which they base their actions is a judgement about the value of particular forms of knowledge and the social identities that are associated with them. This explanation is of particular interest in the context of our work where we anticipated that home and school mathematics would be differentiated by both parents and teachers not only on the basis of the mathematical tools used in them, but also in terms of their social status. Goodnow (1988, 1990, 1993) has argued that a failure to attend to the impact of value judgements about the status of different forms of knowledge has stood in the way of the development of clear accounts of children's learning.

Furthermore, Goodnow (1988) argued that how parents and children negotiate their differences either in knowledge or in positions regarding the social value of an activity depends on 'two-way influences'. In her view:

> 'Parents mark information as having varying degrees of importance. They provide it, withhold it, or frame it on the basis of judgements about value and about the other party's need, age, or capacity to cope. They are selective in the social messages they pass on. Some they endorse, others they subvert or try to exclude from the child's awareness. Such active management is not restricted to parents. We have a great deal yet to learn about the criteria that lead children to decide what information they will transmit, which members of an older generation will be the recipients, and under what circumstances' (pp. 63–64).

Goodnow (1996) suggested that agreement can be reached when each generation takes into account the view of the other. In her opinion this can require that 'each generation monitors where the other stands (a cognitive process) and is interested in resolving differences or reducing any disharmony they may create (a motivational process)'. On the other hand, when either the child or the parent holds on to particular positions or regards them as non-negotiable, divergence will be the consequence. In addition, divergence may also occur when neither the child nor the parent is aware that a difference exists. It is apparent that the way Goodnow described the interactions addressed the double character of the representations mentioned above and the corresponding psychological cognitive and identity processes.

In the next part of this chapter we describe empirical research into the management of the transition between home and school in relation to mathematics. In what ways is children's performance in the subject at school influenced by the negotiation of what Beach (1999) termed collateral transitions – the simultaneous participation in mathematics practices at home and at school (see chapter one). We will in particular try to understand the influence of a key social actor in this process – the parent.

3. THE RESEARCH CONTEXT AND METHODOLOGY

3.1. Children in Multiethnic Schools

Multiethnic primary schools in England seem to be an interesting situation in which to continue our investigations of home and school mathematics. Previous studies of ethnic minority students in this country have highlighted wide gaps between their lives at home and at school (McIntyre, Bhatti, & Fuller, 1997) and problems in the way they are assessed at school (Cline, 1998). Recent surveys also show substantial evidence of an interaction between ethnicity and achievement in mathematics (Gillborn & Gipps, 1996). Finally, patterns of ethnic group participation in the schools are more heterogeneous in this post-colonial situation than in the societies in which most of the research in this tradition has been conducted previously.

Multiethnic schools in this setting have some similar characteristics to the Brazilian schools in the original studies that generated the development of theory, ideas and new questions on relationships between home and school mathematics (Abreu, 1995a, 1995b and 1999). Some of them have on roll children from ethnic groups who, on average, underachieve at school. But it is likely that within any single year group there will be wide variation in performance among the children from the same ethnic group – including both high and low achievers. An important difference from earlier studies is that differences between home and school mathematics are likely to be linked to parents' experiences of a different culture and a different school system through going to school in their country of origin on another continent. In contrast, in the Brazilian study the focus was on parents' experiences with a distinctive non-school mathematics in the same area.

3.2. The Children, the Schools and the Parents

The data presented in this chapter were obtained in a study that consisted of a series of linked case studies. Twenty-four schoolchildren, their teachers and their parents participated in the study. The selection of the children took into account school performance in mathematics (high versus low achievement), ethnicity (White-British versus Pakistani-British[2]) and level of schooling (years 2, 4 and 6). Pupils are 6–7 years old in year 2, 8–9 years old in year 4, and 10–11 years old in year 6.

The children were selected from four schools serving two multiethnic areas of a small industrial town in the South of England. In each of the four schools ethnic

[2] All the children involved in the project were British. For the purpose of distinguishing between the two ethnic groups in the report we use 'White-British' for children whose parents were born in the United Kingdom, spoke English as their first language and came from the majority community and 'Pakistani-British' for the children whose parents came to this country from Pakistan and spoke Urdu as their first language.

minority students comprised well over half of the total school population, with substantial Pakistani representation. Thus a high proportion of the students came from homes where English was not the first language.

One parental interview was conducted for each child. This included 12 White and 12 Pakistani families. The final sample comprised 16 mothers, 2 fathers, 5 couples (mother and father) and one couple of grandparents. (Relatives such as sisters and aunts were often present in interviews with Pakistani parents)

3.3. Interviewing Children and Parents

The main data analysed in this chapter were obtained in interviews with the children and their parents, but this will be complemented with information from other sources relevant to the theme (e.g. field notes, interviews with teachers and classroom observations). The interviews with the children were conducted individually in the school in a private room (not in the classroom). The sections of the interviews that we analyse here refer to their articulation of their views on the help they received at home with their learning of mathematics. An in depth interview was conducted with one or both parents after we had concluded a series of interviews and classrooms observations with the child and his or her classroom teacher. All the data from parents were collected by one of the authors (Tatheer Shamsi). Like the Pakistani children in this project, Tatheer grew up in England in a multicultural and multilingual environment. Her command of the parents' first languages and her close understanding of their cultural background were crucial factors in the conduct of the research. They facilitated the establishment of intersubjectivity and the negotiation and clarification of meaning.

Seven parents chose to be interviewed in their child's school and 17 in their homes. Most interviews with Pakistani parents were conducted in Urdu (4 exclusively in Urdu, 6 in Urdu combined with English, and 2 in English). The design of the parental interview assumed that their social representations of mathematics learning would reflect both their own experiences of a particular school culture and their perceptions of the schooling of their own children, and that this would create a distinct home mathematics culture for the children. On this basis a set of questions was prepared to explore: (i) parents' participation in the child's learning in terms of their involvement at different stages and their role in initiating mathematical activities and in supervising activities at different stages; (ii) parents' own involvement in the activities and how it changed over time. The questions were designed to stimulate parents to talk about concrete experiences and to allow a focused but semi-structured style of interviewing.

3.4. The Researcher's Impressions from Visiting and Interviewing Parents in Their Homes

While we were writing this chapter Tatheer Shamsi reviewed her impressions from the meetings with parents. Her account suggests how her cultural and linguistic background contributed to the success of data collection with the Pakistani families. It is also revealing in terms of the way parents framed the interview in itself. For instance, in the Pakistani homes various family members were present and joined in the interview. Other researchers have experienced the same phenomenon in studies with extended families (Gallimore & Goldenberg, 1993). In the White monolingual families it was framed as a personal exchange restricted to the parents.

I don't remember a difference in the location of houses as there was great variation; some were located near the schools, but they were usually a drive away. The better houses were a reflection of parents' socio-economic background rather than ethnicity. I was usually welcomed by all parents, but more so by the Pakistani parents who felt comfortable with me and took an interest in my background, e.g. where I'm from in Pakistan, my occupation, my educational background. A few of the Pakistani parents invited me to other areas of the house to illustrate how learning takes place, e.g. to an adjoining room where the books were kept. The Pakistani parents were far more likely to talk more enthusiastically about their children's educational routine in the home and their own participation in it. They would often veer off the subject and give detailed general information about their children's learning experiences, which were not part of the interview.

The Pakistani parents were more likely to volunteer additional information, and were happy to elaborate, particularly when they were comfortable with the interview – usually when it was in Urdu. The switching of languages was natural, especially when there were other family members taking part in the interview with whom I was usually communicating in English, e.g. siblings. I would initiate speaking in Urdu with the parents, and they usually preferred it, although the siblings preferred English. However, it was often a mixture, with siblings expressing some points in Urdu, and parents some in English. The Pakistani parent interviews were usually more informal, with the parents happy to compare the target child's performance with that of their other children, often giving long accounts of all their children's aptitudes and abilities. It was clear that the Pakistani parents took pride in their children's achievements and were keen to discuss them. There was often a reluctance to acknowledge their weaknesses, even when the children were in the low achieving groups – although this was not always the case (particularly with the better informed parents).

The English parents were more likely to know where their children were performing poorly, and to be taking action in this regard. The interviews with the English parents were more specific and focused, and they were more likely to be concerned about whether their responses were appropriate and answered the questions adequately. Although in some cases both parents participated in the interview, it was generally noticed, with both ethnic groups, that fathers were reluctant to participate when they felt that mothers were already 'doing a good job.' No English siblings joined in, but many Pakistani siblings as well as other extended family members were present during the interview and participated, e.g. an aunt. This was not due to language difficulties as they were able to understand when I spoke in Urdu with them, but often for support. Sometimes the siblings were more aware and better informed of the child's educational progress in school. The target child was never present during the interview.

The interviews with Pakistani parents were usually significantly longer than the English ones. The Pakistani parents often had an opportunity to elaborate and to reminisce particularly with the questions towards the end regarding their own educational experiences in Pakistan. The length of the interviews varied, but usually fell between 40 minutes and 1 hour 45 minutes. Parents often did not meet with me at the time which was arranged. This sometimes happened with English parents, but was far more salient with the Pakistani parents. And finally, both groups appeared to express equal interest in the aims and outcomes of the project.

4. EMPIRICAL FINDINGS

The findings are described in three steps:

- First the prevailing representations of home and school mathematics among the children and their parents are presented. This will clearly show that both children and parents have come to see differences between the way mathematics is tackled in the child's school and at home. It will also show that faced with the differences children have adopted particular positions, which are visible in their preferences for home or school mathematics.
- Second we will present an overview of patterns of interaction between the child's positioning[3] and the support provided by parents. We will argue that a common thread among high achievers is that their parents have been able to provide support that is sensitive to the particular child's needs and positioning. Conversely, we will show that there is more divergence in the relationships between low achievers and their helpers at home.
- Third we will conclude with three detailed case studies which provide accounts of the transition from the perspective of the child and the parents and explore the relationship with the child's performance in school mathematics.

It is important to note that though the first two results sections focus on 'prevailing' representations they only refer to observations in this study. They are not intended to be generalised to other groups of White-British or Pakistani-British children.

4.1. *Children's Accounts of Their Experiences when Learning Mathematics at Home*

Previous studies (e.g. Abreu, 1995b) had stimulated children to talk about home mathematics by asking them about the activities they usually engaged in after school. We asked similar questions in this project. Children's answers revealed very little engagement with the types of home mathematics linked to the home economy that had been observed in the previous studies in Brazil and Portugal. It was apparent that the mathematics of out-of-school activities, such as shopping, was not a salient feature in these children's lives.

The other set of questions we used in order to explore children's exposure to home mathematics involved asking them whether anyone in the family helped them with school homework. Most children acknowledged having help with maths at home and were able to provide detailed descriptions about who helped them and how. The home mathematical practices salient for the children were those employed by

---

[3] The notion of positioning used in this study is derived from Duveen (1997, in press), and provides a link between social representations and social identities. In this perspective in the process of being exposed to and re-constructing social representations of knowledge a person takes particular individual positions towards these representations – positions that have implications for their sense of personal identity. Like Duveen, we take manifested preference as evidence of a personal positioning.

members of their families to support their school learning. Thus, the two forms of mathematical practice examined were 'school mathematics at school' in contrast to 'school mathematics at home'.

Of the 21 children who received help at home 15 said it was different from the way their teachers taught them at school. We explored what the children in this group thought was the proper or right way of doing it, the easier way of doing it, and the one that would give the correct answer. Overall children manifested a preference for school mathematics (although the other two possible preferences were also represented, indicating some diversity in their positioning).

Justifications for their preferences included cognitive factors, such as 'it is easier to understand at school', linguistic factors, such as 'my mum explains in Urdu', and factors relating to judgements of value and legitimacy, such as 'my mum doesn't know, but teacher does'; or 'mum's methods are old-fashion'. On this basis we were able to describe children's representations of home and school mathematics in terms of:

- ***Knowledge***: Children indicated three different views of the mathematics their parents or other members of the family used when they helped them: it was the same as at school; it was different; or it was sometimes the same and at other times different. These views were not related specifically to the child's ethnic background or age. It was of interest, however, that only high achievers referred to the third view.
- ***Value:*** When children were asked about their preference for advice on the way to tackle a maths problem, the majority of those who saw home maths as different from school maths (9/15) valued the school approach as the 'proper' one. Again, the high achievers were more likely to acknowledge both sources as legitimate (3/4).
- ***Positioning***: Among the children who articulated a clear preference for the place where they found it easier to learn four types of positioning emerged: a declared preference for school; a declared preference for home; a flexible approach in which school was preferred sometimes and home sometimes; and an understanding of the specificity of home and school mathematics. The most common type of positioning was a preference for school mathematics. This is compatible with (and may perhaps be influenced by) a prevalent view within the school system that hardly acknowledges the distinct role that home mathematical practices can play in children's learning.

4.2. Parents' Accounts of their Participation in their Children's Mathematical Education

White and Pakistani parents gave different accounts of how they participated in their children's mathematical education. Firstly, the two groups differed in terms of the mathematical concepts which they thought the family was responsible for teaching. White parents had tended to help their children with counting and simple sums (adding and taking away). Then, as they saw it, the responsibility had been handed over to the school, and their role had become one of helping the child when he or she got stuck. The Pakistani parents rarely mentioned counting. Instead, they emphasised the learning of sums and times tables. Their stress on learning the times tables was rooted in their own experiences of schooling in Pakistan. When developing the teaching of mathematics topics that were not directly linked to work set as homework by the school, Pakistani parents also reported preparing their own resources.

A second difference between the two groups was the involvement of siblings in helping with mathematics at home. They played a major role in Pakistani families for various reasons. One was their ability to cross linguistic barriers. Another was their parents' perception that they were more familiar with the mathematics of the school. In the White group the mother was usually the main helper, and siblings played only a minor role.

A third difference was that the Pakistani parents reported less informal contact with the teachers. This comment was reciprocal. The teachers also reported less knowledge and contact with the Pakistani parents. Various factors seemed to contribute to this, some relating to actual barriers and others to perceived barriers. An obvious real barrier was language. The preference for Urdu among the parents was apparent in the language they chose for the interviews during this research. Some mothers specifically referred to their lack of proficiency in English and explained that they saw this as inhibiting their role in their child's school education. Another issue that emerged was linked to representations parents had about how home-school relationships should operate: some Pakistani parents, in particular, had expectations about this that were at odds with the schools' practices. One final issue that was mentioned by parents in both ethnic groups was the constraints imposed by patterns of work and by the number of young children in a family. Some of the parents found it difficult to get to the schools at the times when teachers were available to meet them.

In the light of the parents' comments about their difficulties over contact with the schools, it will not be surprising that the information about their children's school mathematics that parents believed they had received directly from the schools was limited. They identified their main source of information about the schools' mathematical practices as what their children said about them and what they brought home with them from school. The impression they received was fragmentary. They were often able to tell the researcher what the child was doing at school at that particular moment in time, such as 'take aways' and 'fractions', but they did not have information about the place of mathematics in the National Curriculum. In spite of this fragmentation both White and Pakistani parents had developed quite sophisticated representations of differences between their own and their children's mathematics. Most parents said that in their interactions with their children they were very often confronted with differences between their own ways of tackling mathematics and what the children had learned at school.

Both groups emphasised certain key differences between mathematical practices they understood to be current at their children's schools and mathematical practices with which they were familiar at home. The first point they highlighted was a difference in the content of mathematics and in the strategies used for calculations. For example, some referred to the use of different algorithms for subtraction and division. The second type of difference they highlighted concerned methods of teaching and the tools used in teaching. For example, some emphasised different practices in the learning of multiplication tables, while others stressed the impact of new instruments such as electronic calculators.

In addition, Pakistani parents stressed differences of language. For all three aspects some parents appeared to have developed successful strategies to cope with the differences, while others experienced them as a 'burden'. Having established that differences between the way the child was doing mathematics at school and at home were salient for both the parents and the children, next we look at the interactions between them to try to understand whether some types of interaction were more conducive to success than others.

4.3. Patterns of Interaction between Children's Positioning and their Parents' Practices

The main reasons children gave for their preferences for doing mathematics at home or at school covered the following.

- ***The specific mathematics that should be learned*:** They expressed judgements about the mathematics they encountered at school and at home. These judgements could have a cognitive basis related to the identification of differences in strategies and methods and whether they were seen as facilitating learning. Alternatively, they could have a social comparison basis in terms of how each type of knowledge was seen to be valued.
- ***The mediating role of others:*** They analysed this role in cognitive terms (e.g. commenting on whose explanations they understood more), in affective terms (e.g. commenting on whom they felt more at ease asking for help), or in social comparison terms (e.g. commenting on who was seen as owning knowledge or competence).
- ***The mediating role of language:*** This issue was treated by some Pakistani children at a cognitive level (e.g. commenting on whether they grasped mathematical concepts more easily in one language or the other). For others, however, language preference had an affective-comparative basis (e.g. defined in terms of the language they felt more confident in).

In explaining their preferences the children covered similar aspects of the home-school relationship to those stressed by parents: knowledge of the subject, key social actors, and language. The children's positioning on home and school mathematics viewed in isolation was not, of course, the only source of individual differences in school mathematics performance. But it appeared to be a key factor when viewed in interaction with the type of help provided by their parents as we can see in Table 1. The table summarises key features of the interaction between the child's positioning and their parents' actions. The data suggest that successful interactions are a two-way process where parents can adjust to the child's needs in terms of mathematics

Table 1: Examples of patterns of interaction between children's positioning and their parents' actions

Child's salient aspect of positioning	High achievers	Low achievers
The mathematics that should be learned is the school's	Parents were aware of the differences between their mathematics and the child's school mathematics and developed strategies to support and adjust to the child's preferences which were accepted by the child (Jafar, Y2; John, Y4).	Children indicated resistance to their parents' help based on: (i) not seeing them as competent in school mathematics and mother showing no awareness of the differences (Kashif, Y2); or (ii) seeing mother's knowledge as old fashioned and mother as being dependent on child to show her the difference (Ellen, Y4).
The mediating role of others – with a preference for the home because of the style of interaction	Support was directed towards understanding. Success was seen in time spent (not rushing) (Saira, Y4), or in making it easier (Ellen, Y6).	Child had a preference for the home approach because at school they 'say things once' but had a parent who wanted the child to learn the school way (Waleed, Y4); Mother expressed no confidence in adjusting her methods (Joseph, Y2).
The mediating role of language	Mothers usually helped in Urdu but were very aware that for the child language could be a barrier in understanding. They consciously paid attention to linguistic mediation (Jafar, Y2, Samir, Y2) Mother saw no difference between mathematics in English and Urdu, and they used textbooks in both languages (Saeeda, Y6).	Parents helped in Urdu and had no awareness that the child preferred English (Kashif, Y2); Parents helped in Urdu because they believed that the child would understand quickly in the home language without appreciating that the child was more familiar with the relevant concepts in English (Zahid, Y6). Parents felt unable to help the child with mathematics because they believed that in Urdu it was totally different (Waheeda, Y2; Waleed, Y4)

content, style of interaction or language. In less successful interactions parents do not meet their children's needs either because they are not aware of their preferences or because they are not aware of differences between home and school mathematics, or because they lack the mathematics or linguistic skills or the confidence to adjust to the child's preferences.

To give a deeper insight into the ways in which the children's transition between home and school mathematics was supported by parents we describe some case studies in the next section.

4.4. *Case Studies*

In this section three case studies are reported. Case studies 1 and 3 refer to two boys from Pakistani families. Kashif[4] and Jafar were both Year 2 pupils and were enrolled in the same school and classroom. They were selected by the teacher on the basis of their level of achievement, Kashif as a low and Jafar as a high achiever. The reason we report them here is that what they and their parents had to say illustrates very clearly some of the differences between parents in the ways in which they were supporting their children's transition between home and school. Kashif is an example of divergence between the child's and his parents' representation of the transition and related support. The case of Jafar then illustrates how convergence can be achieved between child and parents. Case study 2 refers to a Year 6 girl from the White group. It was selected to expand the information provided in case study 1. While in case study 1 parents seemed to lack awareness of the way the child experienced the transition, in case study 2 they were aware but seemed to lack the confidence and skills to provide the kind of help required. Case study 2 also illustrates that parents' difficulties in supporting their child's transition were not solely a factor of belonging to a cultural group with a different history and traditions from the mainstream society in which the school is located. It can also result from generational changes within a mainstream cultural group, which exposes its members to 'encompassing transitions' (see chapter 1).

4.4.1. Case study 1: When parents are not aware that transition exposes the child to differences

Kashif's Performance in School Mathematics

Kashif was selected by the teacher as a year 2 pupil who was a low-achiever in mathematics and was from a Pakistani family. She explained her choice in the following way:

'Why did I chose Kashif? He's very, very quiet and lacks a lot of confidence in his maths work. He does maths at home, and dad does a lot of work with him and has someone come in

---

[4] All names have been changed to assure confidentiality.

and do a lot of work with him. But he still doesn't seem to make the progress. Something like subtraction he can handle, he can do that but he just has no confidence in maths at all. He really seems to fall behind.'

Our own classroom observations and interviews with Kashif indicated that the teacher was quite accurate regarding the child's understanding and skill or technique. However, we also concluded that his reading seemed to be the main cause for his difficulties and lack of motivation. He was in fact quite quick at solving the subtractions when they did not involve word problems. The teacher had not mentioned his reading at all. Her position regarding the impact of language was that 'if they've got poor language doesn't mean to say that necessarily they're not going to grasp maths'. In fact, she reinforced her beliefs using the examples of the high achievers in the sample, who were the top students in mathematics, but not in language. At this point it was obvious that the mediating role of language was not seen as the major barrier for his progress in school mathematics.

Kashif's Home Background

From the background interview with the child we learnt that he lives in a bilingual environment. At home he switched between Urdu, to communicate with his parents, and English with his brother and sister. At school he spoke English even when in the playground. With his Pakistani school-friends he also uses Urdu, but made sure to stress 'not at school at play', indicating that speaking in Urdu can sometimes take place in the playground for fun. When asked which language he likes more he promptly replied 'English', explaining that 'It's better. It's easy to say'.

His mother received all her school education in Pakistan. She studied up to Matric (Year 10 – up to the age of 15). The father went to school in Pakistan till he came to England aged 14. He then went to school in England until the age of 16. He passed all GCSEs, apart from mathematics. For him his difficulties arose because he 'didn't know the language' and had 'no one to help'.

Kashif's Account of Home Help

In Kashif's account of the help he received at home with school mathematics the language issue was not mentioned spontaneously. When questioned he confirmed that his mother teaches him in their home language. His replies to the researcher's specific question as illustrated in the extract below were revealing. He experienced the way they teach him at home as being different from the school. This difference was attributed to the mother and sister 'not knowing' while the teacher 'does know' and associated with the superior status of school knowledge, for him objectified in the figure of his teacher.

I[5]: Okay, when your mum helps you does she explain it in the same way or a different way from the teacher?

[5] In the transcription of extracts from interviews the following key is used: I (interviewer); C (child); M (mother); F (father); (...) words or sentences omitted.

C: Different.
I: Different. When your sister helps you does she explain it in the same way as the teacher or different?
C: Different.
I: Different?
C: Yes
I: As well? How is it different?
C: It's because my sister don't know.
I: Your sister doesn't know what?
C: Teacher does.
I: The teacher does know, yes?
C: Yes.
I: What about your mum?
C: My mum don't properly. Don't know what teacher does.
(...)
I: Yes, why is it better at school?
C: Because teacher is important.

Kashif's Parents' Account of their Participation in his Mathematics Learning

Both parents attended the interview, which took place in the child's school (their choice) and was conducted in Urdu. Both parents made clear that since they came back from their last visit to Pakistan (about three months before the interview) they have been helping Kashif with maths. Their help involved teaching basic sums, such as 'adds, take away and times'. They were also trying to influence his self-perception of maths ability. They explained this as:

'We placed [in his mind] that you are clever at maths [okay]. So, you do very good work in maths, and then he works harder'.

In addition they were trying to develop Kashif's awareness of the social significance of mastering school maths.

'He knows now. I have told him that if you don't study, then you will become a refuse collector. Now he says: I won't pick rubbish will I? Can you see? I study everyday.'

It appeared from the interview with his parents that Kashif was receiving conflicting messages. While he could see that he was not doing very well at school, his parents were working hard to convince him that he was clever. They seem to have done this to boost his self-esteem in the face of negative feedback at school. They reported 'Then we found out, and then we did maths work with him last year as well at home, at first we did a little bit with him (...) now this year he is doing more.' They did not mention linguistic barriers at any stage. In fact English language was not a barrier at home as they were teaching the child in Urdu. It was rare for Kashif to have school maths homework. What they did at home was on their own initiative.

Their involvement in helping the child at home also enabled us to explore aspects that were salient in the teacher's and the child's account. Contrary to the teacher they perceived the child as making good progress ('He is quite a lot better than before, he is much better than before'), as clever and quick ('However, he is clever, whatever he sees he can do it quickly') and, finally as quite well-motivated.

In Kashif's account we saw that he experienced home mathematics as being different. For the parents, however, the child's school maths and the way they taught him at home were the same. The mother was convinced she was teaching the child the same maths. In fact, as illustrated in the next extract she saw no difference between the mathematics she herself learnt at school, the mathematics Kashif was learning at school and the mathematics she taught him at home. She also appeared confident in her ability to teach, while her son expressed doubts about her knowledge.

I: Is the maths you are teaching Kashif the same as the maths that you had studied?
M: Yes it's the same.
I: Or is it different?
M: No, no, it's the same, the add, take away and times is the same (...)
I: Have you learned this in the same way?
M: I've learned it in the same way.
I: In Pakistan?
M: In Pakistan. (...)
I: His teacher teaches in the same way?
M: She teaches in the same way (...) There hasn't been any difference so far, maybe when he grows older there will be differences.

To sum up, Kashif is an example where the divergence between the child and the parents was occurring partly because they were not aware of the differences. The parents' representations of primary school mathematics were still associated with their own schooling. These might have been reinforced by their very limited contact with the school, either direct (the teacher reported that she 'haven't really met Kashif's mum') or indirect (since the children 'are not regularly set homework'). The transition for him was associated with an identity, which rejects or undervalues the home ways of knowing and favours school knowledge. His parents seemed to be reinforcing this identity by trying to convince him that decent jobs in the society he lives in require a person to be good at school mathematics. In addition, none of them seemed to be aware that in the transition between learning mathematics at home and at school there was a move from one language to another, which was causing difficulties.

4.4.2. Case study 2: When parents develop awareness of the differences, but experience it as a 'burden'

Rachel's Performance in School Mathematics

Rachel was a year 6 student chosen to participate in the project as a low achiever. The teacher justified her choice by explaining to the researchers that:

'She's in the bottom set for English, and I think she's on the special needs register (...). But definitely in my maths set she's definitely one of the poorest children of a White monolingual, in my maths set. So that's why I've chosen her. But she's very keen, and that's quite unusual for someone of that low ability. Usually, well some children, they get very depressed, and they just kind of like jog along, and they don't really, but she's ever so keen and always wants to try and is really, you know, when she's got something right, she's really happy, and you know, it means a lot to her to be successful. But she does have trouble picking up new ideas, and she is a low achiever in that set.'

Rachel confirmed in the individual interviews that she was keen on learning school mathematics. She said 'I like it. It teaches you … lots of different ways and if you don't have any maths in your life, then you won't get a job'. In the classroom she was also keen to participate, putting her hand up and answering simple questions.

Rachel's Home Background

Rachel had been born and lived all her life in the town where she was attending school. She had two sisters, one older and the other younger than she was. After school on certain days of the week she participated in 'Girl Guides' activities. One day a week a tutor came home to support her with her 'English and spellings'. Her justification for home tuition was 'because we're not allowed to go out of class in year 6 she comes to my house instead'.

Her mother was a trained nursery nurse and worked as a classroom assistant at the Infants school (Four Plus Unit), but she reckoned she knew no more about her child's schooling than any other parent. She was a member of the 'Friends Committee' and carried out fund-raising for the school and also helped at summer and Christmas fairs. Both parents had studied mathematics to GCSE level. Rachel's father reported using a lot of mathematics in his job.

Rachel's Account of Home Help

Rachel's parents did not set her any extra activities at home to support her learning of mathematics or other school subjects. However, they helped her if necessary. As she said:

'When I get home and I'm stuck on something, I ask them to help me'.

She saw no differences between the parents' and the teacher's ways of doing mathematics. In her own words her parents explained:

'The same way as my teacher explains it at school'.

She was not sure whether she understood better at home or at school.

Rachel's Parents Account of their Participation in her Mathematics Learning

Both parents were present at the interview, which was conducted at their home. However, the mother took the lead in answering the questions. They confirmed that, as Rachel had reported, they let her try to work out school homework independently.

Help would generally be provided when she asked. But, as illustrated in the extract below, Rachel's mother seemed to have difficulty in handling this situation. In her view this was linked to her lack of mathematical knowledge. In addition we could see that she tried to 'normalise' Rachel's difficulties comparing with herself: 'I wasn't that brilliant on that'.

I: (...) how would you know whether she needs help then?
M: Well she would come and ask, or sometimes if she's done it all herself, then I'll say, let me look over it and then if I see mistakes I say, you know you've made a mistake, you need to look at it and then see if she can find out where her mistake was, and then if she couldn't then I would pinpoint it out for her and then...go over it.
I: Okay.
M: But I mean some of the maths I don't know so, I have a friend and I just pop across to the friend and she explains it to me and then I come back and relay it to her because my friend's a school teacher.
I: Right, okay. And what sort of maths do you find that she has difficulty with? The areas which she gets stuck...needs your help?
F: Multiplication I think.
M: Yes, I would say multiplication, division.
I: Okay.
M: Probably area, measurement, areas and things like that, but then I wasn't that brilliant on that.

In the sequence of the interview when the mother was recounting her experience of helping Rachel to learn long division it became obvious some of her difficulties were due to differences between the methods of calculation and methods of teaching with which she was familiar and the methods used at Rachel's school. As she exemplified with long division:

'I was trying to show her this way but they don't do it like that at school any more, they do a quick thing, like you'd try and work out your times table'.

These differences were linked to the methods used when she was a child at school. But, also to changes in teaching methods which go in and out of fashion. The way Rachel's mother gained access to knowledge of differences between home and school mathematics occurred as part of an interactive process with her own child. As she stressed:

'I had to ... listen and go her (child) way'.

The experience of listening and learning from the child, however, could be a burden and apparently was stressful for Rachel's mother. She explicitly mentioned her impression that the differences can confuse the child:

'But I was trying to show her that when she was doing it that way she was getting confused '.

Finally, a researcher's question prompted the mother to talk about her worries related to differences between her methods and the way the child was taught at school. As illustrated in the following abstract she felt the amount of help she had been giving was constrained by a worry of not confusing the child and by her not knowing exactly what her daughter was learning at school.

I: Do you worry about helping her, or you know in case that they're being taught a different way?

M: I do get worried, and I've always sort of said, that's why…I want to help her and I suppose that's why I probably don't do as much as with her what I could. Because I don't want to confuse her. … Because I don't know the exact way they're being taught and I've said this all along, I wish there was courses, I know it's hard for schools and that but I wish they would run courses for parents… and be shown exactly how they're taught in school. They need to come home and learn it, and they'll need the same jargon as the teachers use, the same methods because we can only show her the way we were taught and if it's different to how she's taught in school, she'd get frustrated and I'd get frustrated, we'd…just get…at loggerheads. That's why, that and that has happened in the past. So, that's why I suppose now I don't help as much.

In addition to developing a concern about the consequences for the child in terms of cognitive confusion and emotional frustration, she had also become aware that Rachel sometimes manifested resistance to her advice.

'I ask her if she's gone wrong and things like that and sort of show her but I mean if she was adamant that she wanted to leave it like that then I would leave it. (…) And then, or I say, oh you got it wrong, look over it…but I mean if she's not really, I mean they have moods don't they, and if she's not in the mood for it then…just leave it.'

To conclude, what started with a light statement about a possible lack of specific mathematical knowledge was elaborated, as the interview progressed, in terms of the 'burden' that differences between home and school mathematics can bring to parents and children. Besides all the discomfort this can bring to the parents, the most drastic consequences seem to be for the child. As the mother put it these differences can generate a sense of frustration for the child. A clear contrast in Kashif's and Rachel's parents' accounts was their degree of awareness of specific differences between their and the child's school mathematics. It was as if for Kashif's parents there was no transition between home and school mathematics, as if after you master a mathematical concept you could apply it everywhere. In contrast Rachel's parents showed an acute sense of the transition process, but a lack of knowledge and confidence.

4.4.3. Case study 3: When parents develop awareness and strategies to cope with the differences

Jafar's Performance in School Mathematics

Jafar was selected by the teacher as a year 2 pupil who was a high achiever in mathematics and was from a Pakistani family. She described him as a quiet child but confident. In her words 'He's just quietly confident' and 'he knows what to do'.

Again our observations confirmed the teacher's description. He was a fast and quiet worker – far ahead of the others in the table.

Jafar's Home Background

From the background interview we learned that he was born in England, but both his parents came from Pakistan. He reported visiting his parents' home country every year during the summer holidays. He lived in a bilingual environment. At home he spoke Urdu with his parents and English with the brothers. He also often read in Urdu. At school he used English, and occasionally Urdu with Pakistani friends for play purposes (e.g. singing songs). In terms of linguistic preference he said 'I like English better', explaining that that was 'because I've learnt more words in English and less in Urdu'. In addition to these two languages he was also learning Arabic for religious purposes, that is, to read the Qur'an.

Jafar's mother went to school in Pakistan up to the age of 16. She enjoyed mathematics and told the interviewer that her husband was also good at maths. While in Pakistan she worked in a school, first as teacher, then as Head Teacher and finally was promoted to Central Head Mistress. Both parents had had some first hand experience with the English system of education. Jafar's mother did voluntary work as a classroom assistant in local schools. His father did an extra GCSE in mathematics when their oldest son went to high school. This was intended to prepare him to offer his son support at home, but according to his wife that proved to be unnecessary.

Jafar's Account of Home Help

As with Kashif and Rachel, the main home help for Jafar was provided by his mother. Like Kashif he also pointed out differences between his mother's methods and those employed at his school. However, from the extract below we can infer striking contrasts between the experience as reported by Jafar and that reported by Kashif. First, he positively valued the help. In his own words 'she gets it easier'. Second, he could differentiate between instances when the mother did exactly what the teacher would have done and instances when she used a different method. By identifying both similarities and differences he was showing an understanding of the complexity of the situation that was more sophisticated than that of Kashif (who saw it as different) and Rachel (who saw no differences at all). Third, he had meta-awareness of the role played by language. Although he seemed to cope well with maths in Urdu, when not understanding he would ask his mother to switch language. In spite of his fluency in both languages he acknowledged a preference for doing mathematics in English. His account of the help he received at home also sheds some light on why he used mental methods so frequently at school. When comparing the way his mother taught him at home with his teacher's methods, he stressed his mother's oral approach.

I: Is it, what do you like, why do you like it that your mum helps you?
C: Because she gets bit easier otherwise because it's not that easy without help.
I: Okay then. And does your mum explain it as the same way as the teacher?

C: Mmm
I: Or is it in a different way?
C: In the same way.
I: In the same way? So your mum helps, what, the maths that your mum does at home is the same as what your teacher does?
C: Sometimes she does different ones as well.
(...)
I: Okay then. And when your mum is doing it with you which language does she do it in?
C: Urdu and English.
(...)
I: Okay. But, when your mother does it in a different way, she does it differently, you said that sometimes she does it differently, so how does she do it? What does she write? What does she say to you?
C: She doesn't write, she tells me, like sometimes I don't understand it.
I: Yes and then she tells you, but you sometimes don't understand?
C: Mmm
I: Okay, and then what do you say to her?
C: That she says it in English.
I: Okay, then you understand it?
C: Mmm
I: So you, okay, so you understand it when she says it in English?
C: Yes, and Urdu sometimes.
I: And Urdu sometimes, okay. Okay, so which way do you think is the proper way to do it?
C: Well English is easier because I understand it more.

Jafar's Parents' Account of their Participation in his Mathematics Learning

The parent interview was conducted with the mother in Urdu and took place in the child's school. We noted that right from the start the way the mother structured her help at home was based on what the child was going to be doing at school. Thus, when she introduced her child to counting and simple adding and take away sums before he started school, she did not follow a 'home curriculum' independent of what the child would then be exposed to at school. As the following extract shows she perceived the child as sensitive to interactions between home and school and took measures to prevent the child becoming bored or confused by different methods.

I: At first you started at home...
P: He already did these at home, and then when I noted that he wasn't doing it at school and it was boring for him then I stopped it a little.
I: Right.
P: Because when he will start it in class then he will get bored.
I: Right, so you stopped in order to let him learn it in class first.
P: Yes.
I: And then.
P: In case I had a different method...
I: Right.
P: Then the child would have got confused.

Besides being alert to what goes on in the school this mother also showed an understanding of the child's preferences. She described the differences between his motivation in completing homework set by the school and his motivation for similar tasks set at home in the following way:

> 'I've seen that if something is shown at home, the child will take interest, but when he then does the same thing at school, then he becomes more keen.'

A good deal of information about school maths seems to have come from the child, as the mother said:

> 'I would ask every day what he had learned at school'.

She also appears to have been able to get detailed information about school methods:

> 'I had asked him, I had asked what he was doing, I did similar sums with him at home and asked what do you do, and he said that I'm taking away, you borrow that, then I wrote the sum for him, do you do the sums like this, he said yes in this way, then I gave him a few more sums at home for practice, so that you should do them'.

She tried to get more information directly from the school and the next extract is her account of how she struggled to get access to her child's classroom work. One can see she did not get what she really wanted. She was given some information verbally, but access to her child's work was only available during open evenings. The school did not set homework, and they told her it was not part of the school's policy to send exercise books home. So it seemed that she did not have a source other than Jafar himself for day to day information. However, the other point that is clearly illustrated in this extract was that the way that parents support their children was not independent of the way that the school functioned. The school controlled the information that they gave parents access to, even when a parent clearly articulated a request in an apparently unobtrusive way, such as asking to look at the child's school books.

P: After school when I went to collect my child, I said to her (the teacher) that I'd like to see the child like this, in fact I had given Jafar a letter. (...) That I would like to meet with you (...) and discuss the child. Then when I went to collect him I said to the teacher that I would like to see Jafar's work, and then she had said to me that the open evening is coming soon. (...) Then I said to her that I'm actually worried and want to know whether he's working well in class
I: How long before the open evening was your meeting?
P: About one, half a term before (...)
I: So you see reports as well? Which other forms of...
P: The school does not give us any reports. They just give it at the end of the year. (...) But the case is that if you meet with the teacher then she will tell you verbally. (...) But the work is only shown at the open evening.

The above observations suggest that this mother was aware of the child's performance at school. To get this information she used quite sophisticated strategies. One might speculate that her insider knowledge of the system after working in local schools as a teacher's assistant provided her with the resources to ask specific questions both of the teacher and of her child. This knowledge base was then a foundation for developing a strategy for helping the child to experience the transition between home and school as a positive experience. But Rachel's mother was also an insider to some extent and yet was still facing difficulties. So probably the success that Jafar's mother had was also partly due to her competence and confidence with primary mathematics as an ex-teacher.

Finally, when asked about whether the maths that she had learned at school was the same as what she teaches her children, she replied 'To start with it's the same, but further on obviously, the difference I notice is that we can help them up to the younger classes'. She went on articulating this position as related to the languages one has been taught: 'But, because we've studied in the Urdu medium in Pakistan (...) and here it's English medium, so when its higher classes I can't then help them.' This account was very close to the one provided by the child, who indicated that the choice of language can sometimes facilitate and at other times hinder his understanding of mathematics.

5. CONCLUSIONS

As noted above, our intention is not to generalise this case study data, but to seek insight from the analysis into possible features and ways of conceptualising the transitions between practices of mathematics at home and at school. Bearing this in mind we would like to draw the following conclusions from our empirical report:

Firstly, adopting a perspective that considered the double character of the mathematical representations helped to bring to light features of the experience directly associated with properties of mathematical tools (e.g. long division) and to the valorisation (e.g. which form of division is seen as legitimate). Thus, for instance, Rachel's mother when describing differences in her methods of doing division was focusing on the mathematical tool. These differences, however, were not described in neutral terms. The different methods were not treated as equal alternatives. Instead, in the context of helping her daughter to do homework the mother clearly stated she had to listen to Rachel and go her way.

Secondly, like previous researchers in social representations, we found that the differences between home and school mathematics were salient for those who have to face them, i.e. for the parents and the children. The case studies illustrated the positive effects that occur when parents are aware that although for them the mathematics at home and at school can be seen as the same, the child can still experience it differently. In one case described here this realisation enabled the mother to adjust her help at home in a way that was supportive of the child's school success. But, although such awareness is necessary, it is not sufficient in itself if parents lack the required knowledge or confidence to tackle the differences constructively. This, in fact, can lead to parents experiencing the whole process as a 'burden'.

At this point we would like to suggest that the way parents support their child's transition between the home and school mathematics can be conceptualised in terms of: (1) their representations or particular theories of this transition process; (2) the identities they project for their children. In our view this latter feature becomes apparent when one observes that the ways in which parents choose to support the child are not neutral. We can safely say that those who were interviewed were operating on the basis of what they represented as best for their child. Some parents, both White and Pakistani, consciously avoided exposing the child to their own ways of understanding and tackling mathematics. Other parents intentionally engaged the child in their own approach. However, whether parents were sensitive to it or not, we observed that the children did not simply adjust to their parents' expectations.

This led us to believe that the parents' theories of transition that best support the child's school success were continuously re-constructed taking into account the input of the child. This was illustrated in Jafar's case study. Jafar's mother had developed representations of home and school learning, which included a theory of how her child might experience the transition. There was a convergence between the two settings and an awareness of both perspectives in the way that methods, language and identities were negotiated. The child acknowledged the mother's competence and was satisfied that, whilst like the others he accepted the school's as the authoritative point of reference, differences between her practice and that of the school did not necessarily mean that hers automatically had 'inferior status'. The mother acknowledged the child's development of a preference for engaging in learning activities set up by the school and took account of this by carefully adjusting her home teaching to what he was actually doing in his maths lessons at school.

Looking back at our aim of examining home-school transition in order to get insights into individual differences as a whole the case study analysis seems to support the idea that there is one level where individual variation in children's performance in school mathematics can be seen as influenced by features of the transition between home and school which have a sociocultural foundation. In moving between their home and their schools the children are likely to be exposed to different mathematical tools, different media to communicate mathematical ideas, different ideas about how similar tools can be used, and different views on the value of each of these aspects. The impact that any of these might have on the child's developing understanding of mathematics and its learning seems partly to depend on their participation in specific practices in mathematics and mathematics learning. So it depends too on how these practices are represented and structured by the adults in charge (both parents and teachers), not only in isolation, but in relation to each other. This study highlighted the conflicts and resistance that emerge as a consequence of parents' and children's understanding of the process.

REFERENCES

Abreu, G. de (1993). *The relationship between home and school mathematics in a farming community in rural Brazil.* Unpublished Doctoral Dissertation, University of Cambridge, Cambridge – UK.

Abreu, G. de (1995a). *A matemática na vida versus na escola: uma questão de cognição situada ou de identidades sociais* [Mathematics in everyday life versus school: a question of situated cognition or of social identities]. *Psicologia: Teoria e Pesquisa, 11*(2), 85–93.

Abreu, G. de (1995b). Understanding how children experience the relationship between home and school mathematics. *Mind, Culture and Activity: An International Journal, 2*(2), 119–142.

Abreu, G. de (1999). Learning mathematics in and outside school: two views on situated learning. In J. Bliss, R. Säljö & P. Light (Eds.), *Learning sites: social and technological resources for learning* (pp. 17–31). Oxford: Elsevier Science

Beach, K. (1999). Consequential transitions: a sociocultural expedition beyond transfer in education. *Review of Research in Education*, 24, 101–139.

Bernstein, B. (1973). *Class, codes and control.* (Vol. 2). London: Routledge.

Cline, T. (1993). Educational assessment of bilingual pupils: getting the context right. *Educational and Child Psychology, 10* (4), 59–69.

Cline, T. (1998). Assessment of special educational needs for bilingual children. *British Journal of Special Education*, 25, (4), 159–163.

Duveen, G. (1997). Psychological development as a social process. In L. Smith, P. Tomlinson, & J. Dockerel (Eds.), *Piaget, Vygostky and Beyond*. London: Routledge.

Duveen, G. (in press). Representations, identities, resistance. In K. Deaux & G. Philogene (Eds.), *Social representations: introductions and explorations*. Oxford: Blackwell.

Duveen, G., & Lloyd, B. (Eds.). (1990). *Social representations and the development of knowledge*. Cambridge: Cambridge University Press.

Gallimore, R., & Goldenberg, C. (1993). Activity setting of early literacy: home and school factors in children's emergent literacy. In E. A. Forman, N. Minick, & C. A. Stone (Eds.), *Contexts for learning: sociocultural dynamics in children's development* (pp. 315–335). Oxford: Oxford University Press.

Gillborn, D., & Gipps, C. (1996). *Recent research on the achievements of ethnic minority pupils*. London: HMSO.

Goodnow, J. J. (1988). Children, families, and communities: ways of viewing their relationships to each other. In N. Bolger, A. Caspi, G. Downey, & M. Moorehouse (Eds.), *Persons in context: developmental processes* (pp. 50–76). Cambridge: Cambridge University Press.

Goodnow, J.J. (1990). The socialization of cognition: what's involved? In J.W. Stiegler, R.A. Shweder, & G. Herdt (Eds.), *Cultural Psychology* (pp. 259–286). Cambridge: Cambridge University Press.

Goodnow, J.J. (1993). Direction of post-Vygotskian research. In E. Forman, N. Minick, & C.A. Stone (Eds.), *Contexts for learning* (pp. 369–381). Oxford: Oxford University Press.

Goodnow, J.J. (1996, August). The several bases to cross-generation patterns: models and relevant data. Poster presented at the *Biennial Meetings of ISSBD*, Quebec City, Canada.

Heath, S.B. (1983). Ways with words: language, life and work in communities and classrooms. Cambridge: Cambridge University Press.

Lave, J. (1988). *Cognition in practice*. Cambridge: Cambridge University Press.

McIntyre, D., Bhatti, G., & Fuller, M. (1997). Educational experiences of ethnic minority students in Oxford. In B. Cosin & M. Hales (Eds.), *Families, education and social differences* (pp. 197–220). London: Routledge.

Tizard, B., & Hughes, M. (1984). *Young children learning*. London: Fontana Press.

Vygotsky, L. (1978). *Mind in society: the development of higher psychological processes*. Cambridge, Mass: Harvard University Press.

CHAPTER 6

TRANSITIONS BETWEEN HOME AND SCHOOL MATHEMATICS: RAYS OF HOPE AMIDST THE PASSING CLOUDS

MARTA CIVIL AND ROSI ANDRADE
University of Arizona

This chapter draws on a research project[1] that has as an overarching goal the development of teaching innovations to promote students' learning of school mathematics, by building on their knowledge of and experiences with everyday mathematics. We begin by outlining the context for our work, including our interpretation of the concept of transitions and the theoretical framework. We then present a research model that forms the basis of our project and that allows us to explore and address our concept of transitions. We will present and discuss data from different components of the model, while raising questions and dilemmas that we face in our research.

1. OUR CONTEXT

Our work takes place in schools that are in primarily working-class, Mexican-American communities. The children we work with may be recent immigrants from Mexico or Central America, or may have been born in the USA to families that have been living in this country for one generation or more. Many students still have family ties to Mexico and travel there with certain regularity, while others do not. Not all students are bilingual or Spanish monolingual, many are English speakers. There are vast differences between, for example, recent Mexican immigrants whose children were born in Mexico and may have been here for two or three years, or less, and families who have been here for one generation or more (Romo, 1999). Thus, rather than guiding ourselves by general cultural definitions of, for example, Mexican-Americans or Mexican immigrants, we take a dynamic approach to culture as meaning the lived experiences of our students and their families. Our approach

---

[1] Project BRIDGE (Linking home and school: A bridge to the many faces of mathematics) is supported under the Educational Research and Development Centers Program, PR/Award Number R306A60001, as administered by the OERI (U.S. Department of Education). The views expressed here are those of the authors and do not necessarily reflect the views of OERI.

G. de Abreu, A.J. Bishop, and N.C. Presmeg (eds.), Transitions Between Contexts of Mathematical Practices, 149–169.

allows us to go beyond the often simplistic characterization of a certain group (e.g., 'Mexican-Americans') that looks at more tangible aspects, such as type of food, folklore, form of dress, and ignores deeper meanings of cultural practice.

In particular, our work is quick to remind us of the diversity inherent in any cultural group, including socio-economic and political factors, and the need to remain cognizant of the impact of each on the day-to-day experiences of the students in these and other communities, lest we fall to relying on the stereotypes that have remained among the obstacles to creating links between students' families and the school.

Additionally, the context of our work is similar yet unlike that of other border communities. Tucson, Arizona has a very different historical context between Mexican and Anglo, very different from California or Texas, for example, though issues are similar (Sheridan, 1986). There has been a Mexican American community in Tucson that has long fostered bilingual education in the promotion of equality in education and other arenas. The constant influx of Mexican immigration has served to maintain this transborder connection.

2. CONCEPT OF TRANSITIONS

We begin with a brief account of a household visit made to the Alvarez[2] family by a teacher and the first author. The Alvarez are the family of one of this teacher's 10-year old fifth grade students, Alberto. When we visited them, they had been living in the U.S. for about eight months, coming directly from a small Mexican town, some seven hours drive from the U.S.-Mexico border. The family was comprised of five individuals, father, mother, and three boys aged 13, 10 and 6 years. At the time of our visit, the father was at work, so our conversation (which was in Spanish) was with Mrs. Alvarez. The family had moved to the U.S. primarily for economic reasons. Like all families we talk to, Mrs. Alvarez was very interested in her children's education; she expressed concern that in Mexico they were being pushed more in mathematics while here it seemed not so challenging. In this regard, she has kept in touch with relatives in Mexico who have children about the same age as her own, and has been making comparisons of the two educational systems. As an aside, we want to point out that Mrs. Alvarez is not unique in this feeling. Many recent immigrants share the perception, for example, that they find the mathematics instruction more demanding in Mexico than in the U.S. When it comes to computation, many teachers have corroborated that students recently arrived from Mexico, who were attending school there, are quite proficient and are often more familiar and comfortable with arithmetic-type situations than their peers schooled in the U.S.

One of the principal reasons why this teacher had chosen to visit this family was to try to gain a better understanding of who her student Alberto was, because he was not doing too well in school. It turned out that Alberto, while in Mexico, was

---

[2] All names are pseudonyms.

actively involved in the family business, a bakery. He had his own set of customers to whom he sold bread daily after going to school. According to his mother, he managed every detail of his part of the business – the orders and the commercial transactions. Mrs. Alvarez's description was filled with a mother's pride at a son's successful enterprises. In a sense, Alberto's experience is a case of transitions. It is probably an example of both, lateral and collateral transitions (Abreu, Bishop, & Presmeg, chapter 1, this volume). The case of Alberto reflects the transitions involved in a recent immigrant child having to learn how to adapt to a new country, a different school system and another language, who furthermore has left behind friends and his 'business' (both Alberto and his mother told us how much he enjoyed having his own set of customers). Our household visit offered us valuable insight as to why Alberto might not be doing so well in school. Alberto did not want to come to the U.S. and resented his family's decision to do so. His two brothers were happy with the decision and the youngest one started speaking English shortly after his arrival; for the oldest, it was more difficult, but he was trying to learn English. Alberto did not want to be any part of this new life – including school.

Most of the children we work with are not such clear case of transitions. By this we mean that we cannot 'easily' describe them as in the case of a recent immigrant coming to a new country. The transitions may be, to our eye, much subtler. What our students all have in common is that they are members of minority groups who often do not fare well in school. Further, most of them are economically disadvantaged. In addition to the household visits which allow for in-depth conversations with a family, several of the teacher-researchers in the project have held whole class conversations with their students to gain a better understanding of what they do outside of school. These conversations suggest that often children like Alberto occupy active roles in their community and family environment (e.g., language interpreters for their parents; helping at home in the repair of things, in cooking, in taking care of younger siblings). For example, in a fourth grade class (nine-year olds), out of 27 students, all but 3 had chores to do in the house. Furthermore, about 10 of the students described how they helped their parents with their own jobs (at a restaurant; at a hospital; landscaping; evaporative cooler maintenance; apartment cleaning; computer work; gold prospecting; repairing cars; remodeling).

Quite often, though, these experiences and knowledge are not accounted for in their schooling. There seems to be a growing chasm between how these students are perceived and how they perceive themselves and are moved to action or inaction at home versus at school. At home they contribute to the functioning of the household; at school, they often occupy the lower echelons in academic subjects and are cast in a rather passive role. It is in this sense that we view the concept of 'transitions.' For a time now, we have been immersed in shifting contexts between social, cultural and historical pasts with experiences in the present. The effect has made us clearly aware of the implications involved for children and adults alike, in making transitions from a past to a present situation involving pedagogical implications. We have found the need for rays of hope that are forged by a philosophy of strengths based on past knowledge and experiences. It is in this way that we confront the work of shedding light on the connections between home and school mathematics. Our work aims to develop a model that incorporates the multiple transitions related to the mathematics education of minority and working-class background children and their families.

3. THEORETICAL FRAMEWORK

Of particular relevance to our work is the large body of research contrasting out-of-school mathematics and in-school mathematics (Abreu, 1995; Lave, 1988; Nunes, Schliemann, & Carraher, 1993). Many of these studies document how successful and resourceful people are at inventing their own methods of solution to tackle tasks that they see as relevant to their everyday lives. Yet, some of these studies also document a lower performance once a 'similar' task is presented in a school context. The concept of valorisation of these different forms of mathematics (Abreu, 1995) is important to our work as it sheds greater light on our observations of children's performance and perceptions of themselves at school and at home.

Our theoretical approach to learning is grounded in sociocultural theory (Forman & Carr, 1992; Moll, 1992). This approach, while grounded in theory that suggests that culture is dynamic and related to social reality, not folk notions of culture, has the direct intent of applying innovative ideas (based on the knowledge and experiences of students and their families) to educational practice.

In particular, Rogoff's (1994) discussion of three models of teaching and learning (transmission, acquisition and participation) is very helpful toward this discussion on transitions. In the transmission model, knowledge from others is passed on to the learner (adult-centered); in the acquisition model, the learner discovers the knowledge on her or his own (child-centered); in the participation model, the learner participates in a community of learners. Rogoff views learning as transformation of participation and in fact uses this view for the three models (see also Lave, 1996, on 'learning as participation in socially situated practices,' p. 150). In our work, we find the three models present. The first two (transmission and acquisition) characterize school instruction, while the third one (participation) characterizes the children (and their parents') learning outside school and, to a certain extent, the learning in some innovative school contexts that challenge traditional models. Still within the socio-cultural framework, but more specific to mathematics education, we have found the work by Forman (1996) and van Oers (1996) particularly helpful in our thinking and we will return to these studies in the discussion of our work.

Many of the children (and their families) in the classrooms we work in have often been marginalized. By this we mean to say that the lived experiences of these families, whether in this country or for more recent immigrants in their homeland, are tied to socio-political and economic factors that have limited, in great part, the nature and quality of their formal education as well as the opportunities to act as active agents in their own lifecourse. For this reason, we have turned to research on socio-political aspects in mathematics education and implications of critical pedagogy (Frankenstein & Powell, 1994; Secada, 1989; Skovmose, 1994). Secada's

(1989) question 'how should we educate individuals from groups that have suffered discrimination to live in a world in which they are likely to be subjected to similar treatment?' (p. 46) is particularly relevant to our local context. School failure continues to be endemic to minority education (especially for Latino students from economically disadvantaged backgrounds). While multicultural and bilingual education have been responses to these trends, they often continue to replicate those institutional practices that deem minority students to less demanding and engaging educational practices and academic success in great part because they have *not* addressed socio-historical underpinnings of educational reform. We share with Nieto (1999) the belief that a fundamental goal of multicultural education is (should be) 'to promote student learning' (p. xvi), and we share her concern with the fact that 'curiously missing from discussions in most schools that claim to 'do' multicultural education are statements having to do with *student learning*'(italics in original, p. xvi).

4. A RESEARCH MODEL

Our project is classroom-based, with as many as five schools represented at any given time, but with only one to three teachers per school. We work with a small group of teachers (no more than 10) who teach children ranging in age from 8 to 13. The two middle schools and two of the elementary schools serve a working class student population (90% to 95% Latino, primarily of Mexican descent). The third elementary school serves a more mixed population in terms of economics and with 35% to 40% Latino. Two of the schools follow a very specific mathematics curriculum, based on a reform oriented textbook series. The other three schools have a less structured curriculum. Although each teacher (and university researcher) inevitably brings his/her own agenda and set of needs and interests, a point in common in the project is an interest in looking at the mathematics education of the students in these classrooms from a more global point of view, thus trying to bring together the required curriculum and the children's everyday lives. What our project aims to do is to bridge between two different domains – home and school – by building on children's or parents' experiences and knowledge. The sociocultural experiences that form part of these children's worlds provide possible contexts for in-school learning.

There are three key groups of people engaged in this process: the students; their families; the teachers/researchers (including university researchers). Our research model addresses each of these groups through a variety of activities and sources of data as follows:

- Students: interviews; classroom observations.
- Families: household visits; occupational interviews; mother/daughter math club; mathematics workshops for mothers.
- Teachers: interviews; classroom observations; in-class project; participation in study group meetings (with university researchers); presentations at professional meetings.

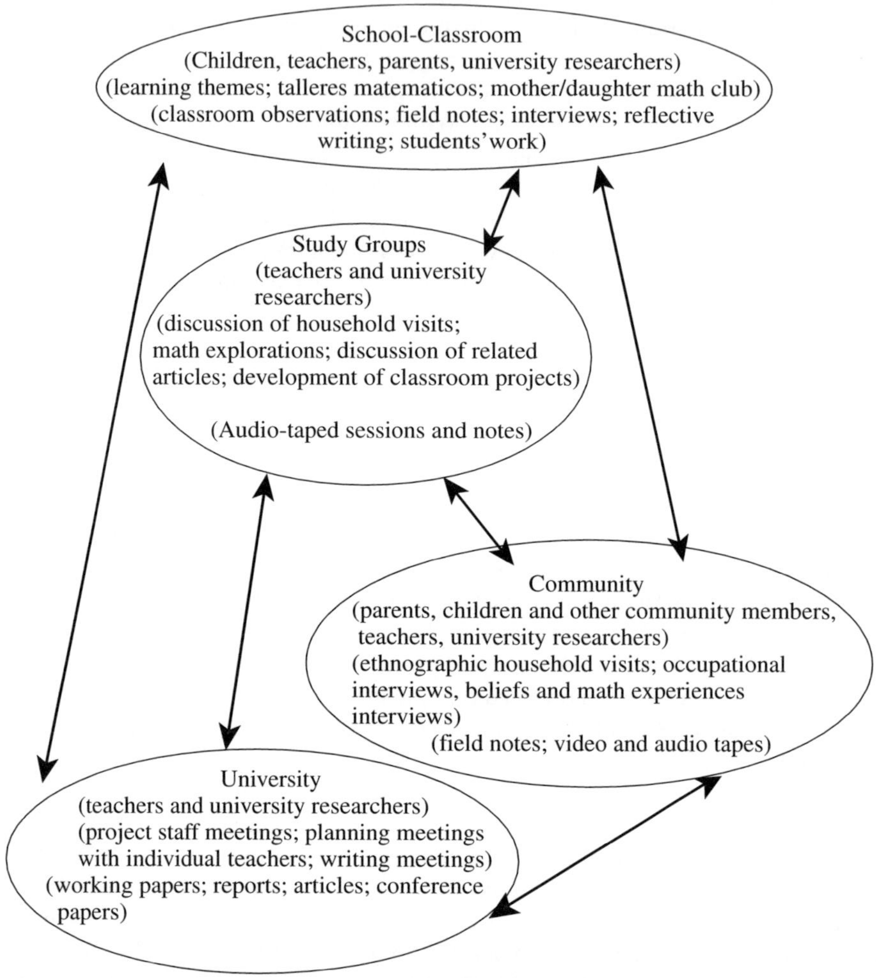

Figure 1. Project BRIDGE Research Model

Figure 1 presents a diagram of the research model for our project. The double arrows represent the two-way bridges that we are establishing between different people, settings, and activities. These bridges also capture the many transitions that all of us in the project experience.

Some of these activities are further developed than others, each bringing another dimension to our understanding. Also, some of them are central to the project. For example, the Study Group meetings (every three to four weeks, at one of the participating schools, where the project teachers and university project staff get together for about two hours) are key to the research process. At these meetings we debrief on the household visits, we engage in explorations of mathematics and we collaborate in the development of learning modules. In what follows, we will look at some of these activities (and related data) from the point of view of their role in helping us address the concept of transitions.

5. HOUSEHOLD VISITS

Early on in the project, teachers receive training in ethnographic research methods to facilitate their household visits. They then visit the homes of some of their students to *learn* about experiences, activities and life in these households. Questionnaires focusing on the family structure, parental attitudes towards child-rearing, labor history, and household activities are used to provide some structure to these home visits. These questionnaires also include a number of questions 'aimed' at uncovering the mathematical potential in the households, which has proven to be no easy task. A key effect of these household visits is on the teachers' perceptions of their students. By seeing and learning about their students' household experiences, teachers begin to develop a firsthand understanding of such experiences, as opposed to accepting often-told generalities about the 'minority culture.'

Teachers also learn about the vast networking practices in the community and about the children's participation in household and community activities. This is where teachers begin to see their students as active participants in the functioning of the household. Additionally, during the course of interviews, teachers find out how the adults in the household have learned their practices and how they engage their children in learning. In most cases, teachers glimpse at living examples of learning by participation in a community of practice. For example, in one of these visits, the father, who has experience in construction, carpentry and car mechanics, commented on how he sometimes takes his children to his work site, and that he also engages them (boys and girls, he pointed out) in projects around the house, such as building a fence, setting a concrete floor, and working on cars in their backyard. But both, mother and father, stressed that the most important goal for their children was that they get a good education (meaning formal / school education). The father made it very clear that he did not want his children to work in construction. He wanted them to graduate from high school and go on to college. He wanted them to be knowledgeable about construction, carpentry, and mechanics because it is useful knowledge, not because he wanted them to go into those fields.

All teachers have commented very positively on the affective impact that these household visits have had on them, as well as on the students whom they have visited. From individual teacher's interviews, we learn the following:

> Based on the one home visit, I know what the student does in her family. And what the family does. It makes me more sensitive to asking questions that I know she [the student] knows the answers to. It is great. She is now participating more in class. (Teacher's interview)
>
> I had known theoretically that yes, there is a lot of information out there. I am not working from a deficit model. But I still needed this first hand experience, even after I read all articles. I know their strengths now. But now that I know that from the home visits, how do I make sure that the child is aware of this, that they have all these experiences? (Teacher's interview)

> I guess realizing that the home is a real learning place, real learning environment, you know I didn't think it was so much a learning environment as it is. (Teacher's interview)

But what about the cognitive impact from the household visits? At a Study Group meeting, one of the teachers, who had recently joined the project, asked 'how do you make the curriculum out of the home visits?' How can we bridge from what we learn in the household visits to the school curriculum? Another teacher pointed out the following:

> As far as [the household visits] impacting my teaching ... first it has to impact my thinking and then at some point it will impact my teaching. (Teacher's interview)

The transformation of household knowledge into pedagogical knowledge for the classroom is not easy. As much as we enjoy the wealth of information that comes out of these household visits, we find ourselves constantly wondering about connections to the teaching of mathematics in school. There seem to be at least two issues involved: one has to do with modes of learning. The household visits repeatedly show children learning through participation in the household practices. Can we bring to the classroom the child's perception of him/herself as a learner outside-school? How can we develop a learning approach to school mathematics that reflects a participation model, rather than the more traditional (in school) transmission or acquisition models (Rogoff, 1994)?

The second issue has to do with content: how can we develop mathematical content out of the household visits? Our goal is not to engage students in the mathematics of construction, or of sewing. Instead we want to look at what are some key ideas that we would like children to explore in mathematics. For example, if the goal is to learn about measurement (linear, area, volume, surface area, etc.) the teacher should know enough about the opportunities that her/his students may have had to engage with these topics outside school in order to bring them to surface in the classroom and help them make connections. But besides knowing about these opportunities, we should explore these mathematical ideas as learners from an approach that is likely to be different from our prior experience when learning mathematics (i.e., by trying to find a way to tie these mathematical ideas to children's activities outside school). This has proven to be a challenge for many of us in the project because of our values and beliefs about what we are willing to count as mathematics. We elaborate on this in the next section.

6. FROM THE HOUSEHOLD TO THE CLASSROOM

A key premise for us is the development of learning environments in the classroom that allow students to participate in activities that are meaningful to them and that at the same time allow them to advance in their learning of academic mathematics. In this sense, we agree with van Oers (1996) characterization of mathematical apprenticeship in the classroom by aiming for activities that are 'recognized as 'real' by the mathematical community of our days' and by immersing the mathematical activity in 'a sociocultural activity that makes sense for the pupils' (p. 106). Thus, one would argue, the sociocultural practices from the household could be developed into academic mathematics activities for the classroom. However, this development is not unproblematic (see also Presmeg, 1998). We face constraints such as the required school curriculum and the amount of time afforded to the development of units or themes, but also we face our own values and beliefs about mathematics, as well as our own experiences with mathematics.

For example, at a Study Group meeting, we had a discussion around whether there is any mathematics in sewing. A teacher, who is also a knowledgeable seamstress was quite reluctant to see much mathematics in sewing. An analysis of our own language as we spoke of, for example, the practice of sewing, was most revealing: 'you do not have to do math'; 'you just measure'; 'is it math or is it common sense?'; 'you trace over another piece of clothing, what's there to measure?' The discussion also addressed the fact that even if we may see mathematics in sewing, what if the seamstress does not see what she does as mathematics? This led us to consider why the seamstress may not see it as mathematics and to the social imposition of what counts as mathematics and who can do mathematics (Harris, 1997). The discussion also took us to address that not all seamstresses are equal. For example, some may be mostly copying patterns, while other may be creating their own. We suspect that the mathematical demands and skills may be different for these two kinds of seamstresses. We proceeded to analyze one of the interviews of a seamstress in terms of the mathematical content of the practice (e.g., angles; from two-dimension [flat patterns] to three-dimension and viceversa; area; estimation) and dispositions (e.g., persistence; enjoyment of challenge) (González, Civil, Andrade, & Fonseca, 1997). During this process, and all throughout this project, we have often been faced with Millroy's (1992) paradox of whether we can see forms of mathematics that may look very different from the kind of mathematics that we learned in classrooms. One of the study group members asked, 'Is there necessarily a relation between the experiences with [everyday] mathematics and school mathematics? If you have too much school mathematics, does it erase our practical mathematics?' These are important questions in our work, as the experience with another seamstress showed. In one of the workshops for mothers (described later in the chapter), we observed a seamstress as she designed a dress (out of paper) for a teacher in the group. Later that day, in her journal, the first author wrote about how easy it seemed for Señora María (the seamstress) to make that dress and how hard it had been for her (Marta), formally trained in mathematics, to follow the process.

> As I listened to Sra. María saying things such as 'waist is 70 cm, let's add 6 because we need 3 for each pleat, and then we divide by 4' and saw her proceed and produce a perfect dress, I thought that the whole process seemed as mesmerizing and mysterious to me as mathematics lectures are to so many people – seeing bits and pieces but not getting the whole picture. [Marta's journal]

On another occasion, a teacher brought up the notion of there being different levels of mathematics, as she referred to the fact that she had met unschooled people in Mexico who were very good at basic mathematics through their work experience. This in turn brought up the questions of, if there are these levels of mathematics, is what we are doing in BRIDGE only going to address the very basic mathematics? Are we not going to get into the more abstract mathematics? This is certainly a concern that several of us share and that we have brought up elsewhere (Civil, 1995; 1998). The issue is how to bring the different forms of mathematics together and still have the students engaged. For example, in a fifth grade classroom (ten year-olds), bringing in 'everyday mathematics' tasks such as looking for and describing tessellations in our local context, opened up the patterns of participation in the classroom. Students were eager to contribute examples and work on determining the repeating 'tile.' But, as soon as we brought in more 'academic mathematics' such as trying to figure out why regular hexagons tessellate but regular pentagons will not, we started losing students in the discussion and were left mostly with the ones who had been labeled by the school system as 'Gifted and Talented' and who were likely to have had prior experiences with this type of mathematical discussions (Civil, in press).

We have been working on connecting everyday and academic mathematics for quite some time now and seem to have made some progress towards this goal (Ayers, Civil, Kahn, & Fonseca, 1998; Civil & Kahn, 2001; Kahn & Civil, 2001). Thus, for example, the garden unit (explained in more detail in the references just mentioned), provided a context for the students to explore mathematical concepts such as measurement and graphing (e.g., height of plants); area (e.g., of the garden); volume (e.g., amount of soil). The mathematical explorations grounded on the garden theme led students and teacher to explore problems such as maximizing area given a fixed amount of fencing, or the need for a scale in graphing the height of a plant. These problems were contextualized in the students' experiences with their gardens, but were then pushed in several directions to expand the inquiry process, while keeping in sight the mathematics curriculum required by the school district. For example, as their plants grew (the children were reluctant to thin them), the need for bigger enclosed gardens arose. How could we obtain more space while using the same amount of chicken wire? This was an authentic problem for the children. They solved it by manipulating the chicken wire until they obtained a shape that gave them more room for their pots than what they previously had. Most of the gardens ended up looking like a circle or a square with round corners. Did they use 'mathematics' in solving this authentic problem? Many people would probably argue that they did not, that all they did was to try different shapes and see which one would work better. But this experience allowed us to design in-class activities to further explore the mathematical problem of finding the shape with largest area given a fixed perimeter. The transformation of these everyday experiences (such as children's need to make their garden enclosures bigger) into pedagogical knowledge for the classroom involves a balancing act between the real world situation and the teacher's mathematical agenda.

This garden unit also served as an opportunity for the development of a new community in the classroom, one that brought together the children, their parents and other family members, the teacher and university-based project members. The various family members contributed their expertise and time with gardening and provided containers, seeds, soil, and other materials needed for the project. The children became very attached to their gardens and developed a special collaborative bond with their garden-group peers. The project was very intense and demanding for the teacher. We cannot even begin to describe how much time, energy, and resources this teacher spent on this theme – developing the garden, going on weekends and holidays to school to water the plants, covering them to protect them from cold weather, going to local gardening resource centers to gather information, etc. And all throughout, she wanted to make sure that she engaged her students in meaningful ('authentic') mathematical experiences. Assessing what mathematics these students learned through the garden theme turned out to be extremely difficult. We had samples of student work, their concept webs around 'math in the garden' and task-based interviews of four children. But we are still not satisfied. Some of these assessment tools are influenced by a cognitive science perspective (e.g., task-based interviews) that, in turn, is more in alignment with traditional assessment in school (where the focus is on individual performance). As Forman (1996) points out 'observing social participation and classroom discourse is rarely viewed as the primary means for evaluating the success of an educational project' (p. 128). Our dilemma, at present, is that although we work towards the development of a community of practice in the classroom (e.g., garden theme) and we embrace the idea that 'learning is a form of participation in the activities of a community of practice and that learning is a discursive activity' (Forman, 1996, p. 128), we are also bound by traditional forms of assessment. The teacher's comment below conveys some of this dilemma as we tried to describe what it is that children were really learning in the garden theme:

> I feel like sometimes I'm limited in my own knowledge as far as what I want to do mathematically. And so, I have to go to books and say, 'now, is that really where I want to go with my 4th and 5th graders? Or do I want to go in that direction? And would this be considered rigorous math? And will it work when my kids get tested on [a district standardized test]? Will they have learned something that will transfer over?' And that's threatening, really threatening. (Teacher's interview)

7. OCCUPATIONAL INTERVIEWS

As we were meeting in the Study Group sessions to discuss the household visits with a mathematical focus, it became clear that as rich as these visits were, it was very hard for us to uncover the mathematics in the household practices. We decided that we needed a series of interviews more focused on practice. To this end, one of the teacher-researchers in the project conducted five such interviews: a mechanic, a carpenter, a welder, a construction worker, and a seamstress. Our main goal for these interviews was to discuss how mathematics was used in these practices. On a limited scale, we wanted to engage in the kind of analysis that Masingila (1994) did with the mathematics of carpet layers, or that Millroy (1992) did with that of carpenters. So, as one goal, we wanted to engage in looking at mathematics in contexts other than academic mathematics to address our values about what we count as mathematics. Since the practices chosen are quite common among the parents of the children in the project classrooms, another goal was to try to establish curricular connections that would enable us to bring in the mathematics from the practices into the classroom. This would also allow us to involve the parents and children in the development of the curriculum as in the garden theme mentioned earlier (cf. Civil, 1993, for an example on a construction theme).

There is yet a third goal for these interviews, a goal that became evident as we started looking at the taped interviews and that is particularly fitting for our purposes in this chapter. It has to do with how the various individuals interviewed reflected on their practice and on how they learned it. It is essentially an affective component and we were struck by the similarity among their comments. In listening to these adults talking about their work and their learning, we were reminded of Lave's (1996) admiration for the Vai and Gola tailors' apprenticeship as a model for learning. The five interviewees described that they had learned their practice by observing first, by paying attention and after a while, by asking questions. For example, in the case of the seamstress, she started when she was six years old by observing a relative sewing and then tried to replicate the same piece of clothing for one of her dolls. Further, all the interviewees demonstrated great pride in what they did and said that they loved their work (and it showed in the passionate way they spoke about it!). The carpenter said that every time he finished a job, he felt proud, 'I did it myself.' The mechanic said 'si vas a hacer una cosa buena, hazlo completo' [if you are going to do something that is good, do it thoroughly]. Similarly, the seamstress recalled the lesson she learned from a teacher she had when she was about 11 years old, 'si vas a hacer algo, hazlo bien' [if you are going to do something, do it well]. The welder, when asked about why he thought that his shop had remained in business for so long (over 30 years), said that his motto was to 'everyday do it better than the previous day; if tomorrow I get a job that is the same as one I got today, do it better.'

As they were asked how they would 'teach' someone who wanted to learn their practice, their replies, again all very similar, reflected a 'learning by participation in the practice'-model (Lave, 1996; Rogoff, 1994). Implicit in their approach to teaching seemed to be that the learner would have to like the practice. The carpenter said this very eloquently when he said that the first thing one would need to become a carpenter was 'las ganas de ser carpintero' [the desire to become a carpenter]. The welder made a similar comment, 'las ganas de trabajar' [the desire to work]. Their approach to 'teaching' varied somewhat, but overall included in this order: first, the desire; second, observation, paying attention; third, learning about the tools and materials; fourth, sequencing of jobs from basic to more complex to help the apprentice learn the trade.

But there were two other key factors in this learning of the trade, that were often repeated by the interviewees; one was the need to have some imagination, to be able to see the project in one's mind. The carpenter said, 'I need to see in my mind the finished piece of furniture.' The other factor was the need to communicate with the customers to understand exactly what they wanted. But it seems that it was more than to communicate, it was to 'get to know your customer very well.' It felt like a very personal connection, as when the seamstress said that before she starts a job, she has to picture in her mind the customer in that dress.

These individuals all emphasized that they had learned by doing, through experience. Interaction with others in their trade was very important when they were learning. When given a job that they were not sure of, they would ask someone more experienced. The mechanic said that when someone brings a motor that he has never seen before, he checks a book as a resource, but he said 'the book helps but it doesn't give you the experience; doing and reading go hand in hand, you need both. The book without the experience is not very helpful.' The seamstress compared her learning experience to that of her daughter who had gone to fashion design school. She said 'school accelerates the process, but experience is very important.' She then remarked that her daughter already had experience at sewing (by working alongside her) before she went to school.

These interviews give us rich examples of learning outside school. They are replete with affective elements such as pride in their work, passion, and enjoyment. Although we did not address how (and whether) they involved their children in their practice, two of them, the mechanic and the seamstress, said they did. Judging by how they all described how they would teach someone who wanted to learn their trade, we wonder about how their children perceive the two experiences – in-school and out-of-school learning. To further understand the transitions that children experience between in-school and out-of-school contexts, and in particular the role that parents (mothers in our case) can play as active supporters of their children's learning of mathematics in school, we started a focus group with Mexican women at one of the project schools. We describe this work in the next section.

8. MOTHERS EXPLORING MATHEMATICS

Through this initial focus group we have been creating opportunities for mothers to come to the school and experience another transition of sorts – mothers attending mathematics workshops designed to develop opportunities for them to engage as learners of mathematics. While these mothers may have given as their initial motivation to come to these workshops their desire to help their children with homework, in time it has become clear that they want to learn mathematics for themselves. They are constantly making references to their own experiences as learners of mathematics when they were in school (in Mexico) as compared to their experiences now. Thus, for example, an area that they often request to explore is algebra. This is due in part to algebra still being a key stumbling block that their children often face in high school, but also to the mothers' own curiosity as they bring back memories about themselves trying to learn algebra when they were in school. These workshops present a series of dilemmas for all of us involved. For one thing, the workshops are centered around typical reform-based mathematics activities, thus engaging the participants in experiences very different from their own as students in school and in some cases even different from what their own children are doing in school. As the mothers compare their experiences in Mexico with their experiences in the U.S., we have to be careful to stress that there is a time factor (things are changing everywhere, not only in the U.S.) and that even in the U.S., not everyone is experiencing a discussion-approach to mathematics, where learners bring in their ideas, and where the emphasis is on making meaning, not on memorizing algorithms. To make things more confusing, these mothers are puzzled by the fact that their relatives' and friends' children who live in Mexico are more 'advanced' in mathematics than their children in the U.S. 'Why is that?' they ask us. The situation is indeed complicated: we face time transitions (mothers as learners when they were young and now) combined with geographical transitions (between regions in Mexico and the Southwestern U.S.) and transitions across different forms of mathematics. In fact these different forms of mathematics bring us to another aspect of our work with the mothers. Not only do we want to engage them in explorations of 'mathematicians' mathematics' through reform-based activities, but we are also heavily influenced by a sociocultural orientation and its implications toward mathematics learning. As part of this orientation and of our goal to establish an authentic dialogue between home mathematics and school mathematics, we want the mothers' experiences with mathematics to become salient in our work. For example, let us consider the task of drawing a quarter of a circle (or a whole circle, for that matter). As a school mathematics task, it usually involves using a compass, deciding on a point as the center and then either estimating when a quarter of a circle has been drawn ('eye-balling it'), or maybe using a protractor or some other way to ensure we have a right angle and then the arc between will be a quarter of a circle. Now, let us look at one of the mothers in this group (Señora María, to whom we have referred earlier) as she explains how to make a dress using paper and real measurements (she used one of the teachers as her model). At one point, she said that she needed to draw a quarter of a circle. Her paper was folded as a rectangle. She then held one end of her measuring tape at one corner of the rectangle (the center of the circle) and marked several points all 25 cm from that corner. She then joined the points and obtained a quarter of a circle (actually, when cutting the paper that would give her half a circle, which is what she wanted). Her procedure of drawing the quarter of a circle shows the circle as the geometric locus of points equidistant from a given point. How can we use experiences such as these to establish bridges between school mathematics and 'home mathematics'?

In what we have found might otherwise be passive forms of participation, mothers and university researchers are collaboratively designing transitions for parents from home to the school, like that of teachers from school to the home. These opportunities take the shape of mathematics workshops (Talleres Matemáticos), where we facilitate and provide resources to support mothers as active learners first, and as subsequent teachers at home. This runs counter to many approaches to parent participation, which all too often cast parents as obstacles to students' academic development by instructing them what to do in fool-proof formulas (e.g., read to your children at least 15 minutes a day) which do not question why or how in meaningful and respectful ways. These forms of transitions challenge our conceptions of how learning takes place and as the mothers have concluded in a reflection on the workshops,

Engaging with our children in the mathematics, allows us to see them [our children] differently, that it is not sufficient to attend to all their other needs, but that it is important that we as parents have these types of discussions. We also realize that though we may not have a certificate in hand, we are also teachers.

In our work, mothers are taking active roles, including presenting at the 1999 annual meetings of the American Educational Research Association. One mother, Emilia Rojo, presented the reflections on the process as prepared by herself with other mothers. They write as follows:

Already in our households, we practice what we have learned in the workshops and our children tell us that it is entertaining. Then we tell them to take notice that the mathematics is not only numbers, subtraction, division, addition, or multiplication, it is also entertaining. Utilizing the different strategies that we are accustomed to using, children do not realize that when they least expect it, they have already subtracted, added, multiplied, and divided or utilized logic.

In concluding on the impact of the workshops, they add:

We believe that we must create a greater consciousness in all parents so that we may become more involved in the education of our children, so that they [our children] may have a better future, not as complicated as the mathematics appears to them.

The mothers take their experiences from the workshops into their homes, and their reflective writing captures the participation model of learning (Rogoff, 1994), as they share the activities with their families:

I am so happy with all these mathematics workshops because I realize how to help my children understand mathematics in a different way, from a fun approach, *all together as a family*. (emphasis added) [BRIDGE, February 2000]

In our work we are trying to develop a holistic model for the mathematics education of economically disadvantaged, ethnic minority children by exploring the different transitions that take place and the key people involved in these transitions – parents (and other family members), teachers, and of course the children themselves[3]. In the next section we turn our attention to them. But we close this section with the writing from a 15-year old – the son of one of the women in these workshops – that once again underscores the participation learning model:

She [his mother] shares it with the entire family and we all get involved in a mathematical reunion that is fun. We are all teachers and students at the same time, there is no difference and that there be much respect and confianza [trust] is most important. [BRIDGE, Spring 2000 newsletter]

---

[3] We acknowledge that there are other key people involved, such as community members and other staff personnel, for example counselors. However in our work we have chosen to capitalize on whom we view as most influential, the parents and the teachers

9. CHILDREN'S IMAGES OF MATHEMATICS

One recurrent theme in our work (shared also by Presmeg,1998) is the question of what counts as mathematics. In the Study Group meetings, we often explore our own views about what we see as being mathematics. The household visits and occupational interviews give us other arenas for this exploration, not only of our views but also of those of the adults involved (e.g., the parents; the mechanic, the seamstress). By expanding our views about mathematics, we aim to develop learning environments in the classroom that encompass different meanings of mathematics, and in this way allow us to further understand the transitions between the different contexts. To this end, it is important that we try to understand how the children themselves view mathematics. We have gathered some data in this respect, but we are still working on how to better access the children's views of mathematics. So far, the questionnaires that we have used include questions such as 'how would you explain what math is to a child younger than you?'; 'what are some things that you like / you don't like about your math class?'; 'if you were the teacher, how would you teach math?'; 'do your parents help you with your math homework? in what ways?'; 'in what ways do you use math outside the classroom?'; 'how do you think your parents use math at home/at work?'

As a rough, first analysis, the children's answers in terms of where they see mathematics fall into three categories. Most children describe mathematics as numbers, counting, and in terms of everyday life, counting money and paying bills. For example, a 12-year old, whose father fixes houses, wrote that his father uses math by counting the money he gets for his work. Yet, the child did not seem aware of other possible uses of mathematics in fixing houses. Some students elaborated on their use of the term counting by saying that they counted certain things (e.g., the dishes they wash) 'for fun.' Others count things for documentation, such as the child who said that he counted the times he took out the trash to then be able to say 'Mom, I've already done this FIVE times!' Allowing students to share their views in a group can provide some further insights as they listen to each other. For example, when one child (8 years old) said that he did not use math in a series of board games that he had just mentioned, another child replied 'yes, you do, like when you count the spaces to move.' But the first child replied 'but that's not math.' Unfortunately, the researcher did not probe further into this, but once again, the beliefs about what we count as being mathematics come into play.

The second category of answers encompasses those who say that mathematics is essential and that it is everywhere. We need to probe further into these answers to gain a better understanding of whether this is a real belief, or an 'imposed' one. The third category, with not that many responses, includes those that seem to have a broader definition of mathematics, such as spatial visualization (as in 'you need math in driving to know how to get to places, when to turn right or turn left'), logical thinking (as in 'understanding and following the rules of a game'). This category also includes those children who showed an awareness for how their parents use mathematics in ways other than counting money and paying bills. For example, one child whose father works as an upholster, wrote that he needed to use math in his job 'to measure the fabric for the chair.' We have some indication that the children who were able to describe wider uses of mathematics by their parents, were those who said that they helped them with their work.

The children's views about school mathematics are another important component in trying to understand the transitions between their different experiences with mathematics. For example, in a focus-interview with a small group of eight-year-olds, some of the things that they did not like about their math class were as follows:

B. : when we have to do thirty problems. You have to think about it a long time and then when your teacher says 'put your stuff away,' it comes to you.
Br.: I don't like to explain how I got the answer since you have to write a lot to do that.
A.: I hate math because I don't like low numbers, I like high numbers. They're harder and you get smarter. With low numbers, it's too easy.

Going along with this idea of wanting harder things to do in math, B. said:

B.: More really hard ones, because it's more fun to learn harder things. A lot of the stuff we already learned it in second grade.

When commenting on how his parents helped him at home with his homework, B. said:

B.: I finish all the homework and they check it and if I do it wrong, I keep doing it until I do it right. They won't tell me the answer, but they explain it on my fingers.

It is interesting to learn more about B.'s thinking about 'telling answers.' Earlier in the conversation, to the question of 'if you were the teacher, how would you teach math?', a student said that he would give them the answers. To this, B. replied:

B.: But they don't learn that way. It would look like they had good grades but they wouldn't be learning if you do that for them.

This student seems to be taking learning as much more profound than obtaining a 'good' letter grade. When asked to think about something he is really good at and what made him good at that, he said:

B.: My bike, because I make mistakes and then I get a lot better.

Just these few answers give us an image of B. as someone who likes to think about mathematics, wants to be challenged, is persistent, and seems to have an approach to learning that includes viewing mistakes as part of the process and making sure that he is learning the material himself not just being given answers.

Another approach that we have used to access children's views of mathematics is through the use of metaphors. We will conclude this section with two examples of answers to 'if math were a flower, what flower would it be and why?'

By a nine-year old, in an interview:

J: It would be a rose because a rose spreads its petals, takes its time, and when it's full it's like when you've finished your math (university).

I: What about the thorns?
J: Things that are difficult, for example, this thorn would be place value multiplication.

By a ten-year old, in writing:

I picked this flower because this flower has a lot of petals and to me the petals are numbers and math has a lot of numbers and this flower is big and so is math with many problems to solve and a lot of difficulties.
I drew the stem because it is the people who do the math and do all the hard problems. I think that the stem are the people that hold the math together.

We hope that through the work described in this chapter we can have more individuals identifying themselves as part of the 'people who do the math' and 'the people that hold the math together.'

10. CONCLUSION

The overall goal of our work is to bridge home-mathematics and school-mathematics. In doing this, we envision several bridges, all of which are two-way traffic (cf. Cole, 1998, for a similar image). Thus, our goal is not to 'take' the home-mathematics and bring it into the classroom, transform it in one way or another and make it 'conform' to the school-mathematics. This is not to say that we do not do some of that, but that is not our ultimate goal, for a variety of reasons: a need to follow a specific curriculum and textbook; the development of learning themes such as the garden theme is very difficult and time demanding; the pedagogical transformation of home mathematics into school mathematics presents several challenges (e.g., risk of 'watering down the school math'; using home mathematics as an artificial, even fake context; trivializing the home mathematics). But more importantly, if our goal were to take the home mathematics and 'transform' it into school mathematics, we might be missing the point, namely the richness in the diversity of aims and values behind the different forms of mathematics

Our aim is to take a more holistic approach to a very complex situation. The bridges represent the many transitions that we (children, parents, teachers, and university-researchers) are constantly going through in the project. For example, while we discuss the transitions involved for immigrant and minority children, we also glimpse another example involving the transition for the teacher in moving from the school to the home. The teachers in this project have all commented how powerful it has been for them to see, often for the first time, their students and their families in a new light. Even the most reluctant and skeptical teachers have commented on the power of home visits:

At first I thought it [going on a home visit] was very stinky. And then once I did the home visit, I was very impressed. … [The student] turned around from a 'C' and 'B' student to an 'A' student. … I was really shocked at the bond that was created. I never anticipated that nor did I anticipate the reception at the home. They were like 'come in, come in!' … I think it is really powerful. I actually didn't want to do the interview. Because that was one of the requirements, I decided I would do it because I had to do it, but I didn't want to do it at all. I was very, very surprised at the results. [Teacher's interview]

The occupational interviews and the home visits provide us with concrete examples of different forms of mathematics (which, as discussed earlier, may actually be quite hard to uncover) and thus provide us with an array of different ways of reasoning (Nunes, 1999). In discussing the findings from the household, or the examples from the occupational interviews, we bring up our values about mathematics. What do we count as being valid mathematics? How can we develop mathematical content out of the household visits? How can we 'uncover' the mathematics in contexts in which we may have no experience or knowledge of, and may be very different from our background in academic mathematics? How do we address the fact that 'only some cultural practices that we are exposed to are treated by the school system as worth knowing' (Nunes, 1999, p. 50)? Reflecting and talking about these issues are important if we are to make sense of the experiences that students may have with different forms of mathematics.

I think that people think that math at school and math in the home are separate things. I know they're trying to make a connection but I don't think in general people even know where the math is. [Teacher's interview]

The household visits and the occupational interviews also provide us with examples of different ways to learn. In particular these examples highlight the importance of affect in the learning process. How can we develop a learning approach to school mathematics that reflects a participation model, rather than the more traditional (in school) transmission or acquisition models (Rogoff, 1994)?

An overarching question in our research is: how can we work with the many transitions that children, their families, teachers and university researchers experience between home and school as an asset towards the mathematical education of these students? In the second part of our title for this chapter, we use the metaphor 'rays of hope amidst the passing clouds' to refer to how home and school may at times overshadow one another, not allowing for lucidity in terms of how one and the other might well support the student.

If we step back from another vantage point, we might well see those 'rays of hope' illuminate a unity between the two and a more hopeful future for students' learning of mathematics.

REFERENCES

Abreu, G. de (1995). Understanding how children experience the relationship between home and school mathematics. *Mind, Culture, and Activity, 2*, 119–142.

Ayers, M., Civil, M., Fonseca, J.D., & Kahn, L.H. (1998). Connecting students' everyday mathematics and school mathematics. In S. Berenson, K. Dawkins, M. Blanton, et al. (Eds.), *Proceedings of the Twentieth Annual Conference of the North American Chapter of the International Group for the Psychology of Mathematics Education* (Vol 2, pp. 533–540). Columbus, OH: ERIC.

Civil, M. (1993). Household visits and teachers' study groups: Integrating mathematics to a sociocultural approach to instruction. In J.R. Becker & B.J. Pence (Eds.) *Proceedings of the Fifteenth Annual Conference of the North American Chapter of the International Group for the Psychology of Mathematics Education* (Vol 2, pp. 49–55). Pacific Grove, CA: San José State University.

Civil, M. (1995, April). *Bringing the mathematics to the foreground.* Paper presented at the Annual Meeting of the American Educational Research Association, San Francisco, CA.

Civil, M. (1998, April). *Bridging in-school mathematics and out-of-school mathematics.* Paper presented at the Annual Meeting of the American Educational Research Association, San Diego, CA.

Civil, M. & Kahn, L.H. (2001). Exploring big ideas in mathematics through a garden project. *Teaching Children Mathematics*, 7(7), 400–405.

Civil, M. (in press). Everyday mathematics, 'Mathematicians' mathematics,' and school mathematics: can we bring them together? In M. Brenner and J. Moschkovich (Eds.), *Everyday and academic mathematics in the classroom.* Journal of Research in Mathematics Education Monograph 11.

Cole, M. (1998). Can cultural psychology help us think about diversity? *Mind, Culture, and Activity, 5*(4), 291–304.

Forman, E.A., (1996). Learning mathematics as participation in classroom practice: Implications of sociocultural theory for educational reform. In L. Steffe and P. Nesher (Eds.), *Theories of mathematical learning* (pp. 115–130). Mahwah, NJ: Lawrence Erlbaum.

Forman, E.A., & Carr, N. (1992, April). *Using peer collaboration to foster scientific thinking: What determines 'success'?.* Paper presented at the Annual Meeting of the American Educational Research Association, San Francisco, CA.

Frankenstein, M. & Powell, A. (1994). Toward liberatory mathematics: Paulo Freire's epistemology and ethnomathematics. In P.L. McLaren & C. Lankshear (Eds.), *Politics of liberation: Paths from Freire* (pp. 74–99). New York: Routledge.

González, N., Civil, M., Andrade, R. & Fonseca, J.D. (1997, March). *A bridge to the many faces of mathematics: Exploring the household mathematical experiences of bilingual students.* Paper presented at the Annual Meeting of the American Educational Research Association, Chicago, IL.

Harris, M. (1997). An example of traditional women's work as a mathematics resource. In A.B. Powell & M. Frankenstein (Eds.), *Ethnomathematics: Challenging eurocentrism in mathematics education* (pp. 215–222). Albany, NY: SUNY.

Kahn, L.H. & Civil, M. (2001). Unearthing the mathematics of a classroom garden. In E. McIntyre, A. Rosebery, and N. González (Eds.), *Classroom diversity: Connecting curriculum to students' lives* (pp. 37–50). Portsmouth, NH: Heinemann.

Lave, J. (1988). *Cognition in practice: Mind, mathematics, and culture in everyday life.* New York: Cambridge University Press.

Lave, J. (1996). Teaching, as learning, in practice. *Mind, Culture, and Activity, 3*, 149–164.

Masingila, J. (1994). Mathematics practice in carpet laying. *Anthropology & Education Quarterly, 25*, 430–462.

Millroy, W. (1992). An ethnographic study of the mathematical ideas of a group of carpenters. *Journal for Research in Mathematics Education*, Monograph number 5.

Moll, L. (1992). Bilingual classroom studies and community analysis. *Educational Researcher, 21*(2), 20–24.

Nieto, S. (1999). *The light in their eyes: Creating multicultural learning communities.* New York: Teachers College Press.

Nunes, T. (1999). Mathematics learning as the socialization of the mind. *Mind, Culture, and Activity, 6*(1), 33–52.

Nunes, T., Schliemann, A., & Carraher, D. (1993). *Street mathematics and school mathematics.* New York: Cambridge University Press.

Presmeg, N.C. (1998). Ethnomathematics in teacher education. *Journal of Mathematics Teacher Education, 1*(3), 317–339.

Rogoff, B. (1994). Developing understanding of the idea of communities of learners. *Mind, Culture and Activity, 1*, 209–229.

Romo, H.D. (1999). *Reaching out: Best practices for educating Mexican-origin children and youth.* Charleston, WV: ERIC Clearinghouse on Rural Education and Small Schools.

Secada, W. (1989). Agenda setting, enlightened self-interest, and equity in mathematics education. *Peabody Journal of Education*, *66*, 22–56.

Sheridan, T.E. (1986). *Los Tucsonenses: The Mexican Community in Tucson, 1854–1941*. Tucson, AZ: The University of Arizona Press.

Skovmose, O. (1994). Towards a critical mathematics education. *Educational Studies in Mathematics*, *27*, 35–57.

van Oers, B. (1996). Learning mathematics as a meaningful activity. In L. Steffe and P. Nesher (Eds.), *Theories of mathematical learning* (pp. 91–113). Mahwah, NJ: Lawrence Erlbaum.

EDITORS' INTERLUDE

THEORETICAL ORIENTATIONS TO TRANSITIONS

The preceding five chapters may each stand alone. Each describes a research project that concerns some of the issues and the dynamics involved in transitions that learners might experience between various mathematical practices. In all of these projects the learning of mathematics is the context or backdrop against which the transitions are examined. Originally it was intended that the learning of mathematics would be the central focus, and not the backdrop, for the concerns addressed in this book. However, it became apparent as the reporting and analyses of the projects progressed, that the complexity of the sociocultural issues involved in these transitions, and the uncharted nature of the territory we were exploring, would be better served if the transitions themselves were the prime focus.

The learning of mathematics is still the prime context for this book, and the analyses and conclusions may have strong implications for mathematics education in any situations where transitions are involved. But it is the boundaries between mathematical practices, and the dynamic processes and links connecting them, which needed to be examined in depth, for the purpose of understanding more about the psychological and sociocultural aspects of these transitions.

In a sense, then, as advocated by Evans (1999), this book is about analysing and theorising the spaces between mathematical practices, for instance at home and at school, or in different countries' mathematics education practices, or in the shifting relations and mathematical understandings within one practice, namely, the selling of newspapers. We have called chapters 2 to 6 'empirical' because data collection took place in each of the research projects reported. However, of necessity each involved theories as lenses through which the transition experiences could be examined. Some of the common elements in these theoretical examinations suggested that further theoretical development of these interpretations is possible. Common themes included the individual construction of identities and values during transitions, as well as the discourses and meanings that were associated with events and processes. Hence the following chapters build on insights from the empirical chapters using various theoretical approaches that address these common elements.

Chapters 2 through 6 related to the first goal of this book, namely, to offer relevant samples of empirical work that help to identify features and dynamics of the experience of transition in different contexts of mathematical practice. The following three theory-oriented chapters address the second goal, namely, offering significant theoretical reflections and accounts of the phenomenon of transitions, from a sociocultural perspective.

G. de Abreu, A.J. Bishop, and N.C. Presmeg (eds.), Transitions Between Contexts of Mathematical Practices, 171–172.

REFERENCE

Evans, J. (1999). Building bridges: Reflections on the problem of transfer of learning in mathematics. *Educational Studies in Mathematics, 39,* 23–44.

CHAPTER 7

TOWARDS A CULTURAL PSYCHOLOGY PERSPECTIVE ON TRANSITIONS BETWEEN CONTEXTS OF MATHEMATICAL PRACTICES

GUIDA DE ABREU
Department of Psychology, University of Luton

1. INTRODUCTION

The empirical studies described in the previous chapters have illustrated the complexity of the experience of learners moving between contexts of mathematical practices. Instead of treating that movement as secondary we have turned it into the primary object of our analysis – thus unifying the focus on the experience of transition. The need for the development of conceptual and methodological tools to expand the understanding of this phenomenon is apparent. This chapter will explore links between key issues currently debated in cultural psychology and issues specific to understanding the experience of transitions in mathematics learning, as illustrated by the evidence in the previous chapters.

It is now acknowledged that a network of several approaches towards a cultural psychology has been emerging during the last century (Cole, 1996; Ratner, 1999). Valsiner (2000) opted for the plural form and suggested that 'Cultural psychologies are new directions in the psychology of the 1990s that attempt to make sense of the ways in which culture assists the person in the construction of his or her psychological world' (p. 1).

Three predominant 'approaches' of cultural psychology were identified by Ratner (1999) differing in the notion of culture used to guide the research. The symbolic approach, which is associated with scholars such as Shweder and Geertz, emphasises culture as 'shared symbols, concepts, meanings, and linguistic terms' (Ratner, 1999, p. 7). The activity theory approach, exemplified through the work of Cole, Rogoff, and other followers of Vygotsky and the Russian tradition, seeks to understand culture as practical cultural activities. The individual approach 'emphasises individual construction of psychological functions from collective symbols and artifacts' (Ratner, 1999, p. 7). This last approach associated with the work of Valsiner shares a notion of culture with the symbolic approach, but emphasises individual agency in psychological development.

G. de Abreu, A.J. Bishop, and N.C. Presmeg (eds.), Transitions Between Contexts of Mathematical Practices, 173–192.

Although the various approaches can be seen as complementing each other this chapter starts by drawing on the activity theory approach, and more specifically on Cole's (1996, 1998) version of cultural psychology, which he calls CHAT (cultural-historical activity theory). The reason for structuring this chapter along these lines is because it has been the predominant approach in the study of development of mathematical thinking and cognition across and within cultures, and is one that has substantially informed my own investigations.

The next part of this chapter is divided into three sections. The first section draws on previous research on culture and mathematical cognition to outline how the concepts of 'cultural practices' and 'mediation' became central in this field of research. The next section outlines how a focus on transitions between practices led the authors of this book to explore the value-oriented aspects of mediation. Special attention is given to the fact that a focus on transitions foregrounds the importance of the valorisation of specific practices by societies, institutions and communities of practice. The final section elaborates the view that the re-construction of valorisation at the personal level is a process of identity construction and positioning. It starts with a parallel between the psychological construction of cultural tools and social value, and proposes that both have an existence on the social plane and will be re-constructed on the personal plane. The concepts of 'identity chaining' (Valsiner, 2000), 'competing identities' and 'projected identities' are introduced to explore psychological issues related to the impact of valorisation of mathematical practices on the person.

2. THEORETICAL BACKGROUND

2.1. *The Shift from Cross-Cultural Studies of Mathematical Cognition to Studies of uses and Learning of Mathematics in 'Cultural Practices'*

Studies of mathematical cognition across cultures and practices form one of the building-blocks that contributed to current conceptualisations of cultural psychology. Gay and Cole's (1967) studies of mathematical cognition in West Africa can be counted as one of the seminal contributions to the discipline. Cole (1995) recounted the influence of this early work in the following terms:

> 'My involvement in the study of culture and human psychological processes began 30 years ago when I was sent as a consultant to John Gay, then a missionary teacher of mathematics at a small college in the interior of Liberia, West Africa. The task: to figure out why Liberian children seemed to experience so much difficulty learning mathematics. My graduate training was in the tradition of American mathematical learning theory, which at that time entailed the use of algebra and probability theory to provide a foundation for discovery of presumably universals laws of learning. I knew almost nothing about the teaching and learning of mathematics, and even less about Liberia'. (p. 25)

Failed attempts to apply standard psychological knowledge acquired in the USA to understand the learning and thinking of the Liberian population led Cole and his colleagues to rethink the role of culture in human development. To cut a long story short one can say that they pioneered the first wave of cross-cultural studies highlighting the inadequacy of Piagetian universal stages to describe the mathematical thinking of adults in non-western cultures (Gay & Cole, 1967). This was followed by a second wave of research drawing on anthropological and Vygotskian theory, which provided detailed accounts of non-western and out-of-school forms of mathematical knowledge (Lave, 1977; Nunes, Schliemann & Carraher, 1993; Saxe, 1982).

In the second wave of studies the shift from cross-cultural comparisons to studies of **cultural practices** was noticeable. Researchers realised that a comparison of how individuals in two different countries solved a similar problem or research task, which was the typical procedure used in cross-cultural studies, did not analyse culture at all (Ratner, 2000; Valsiner, 2000). This procedure simply attributed differences or similarities to the fact that the individuals belonged to a particular group. As Cole and his colleagues demonstrated, without an analysis of the nature of the Liberian farmers' practices the differences could easily be attributed to cognitive processes 'inside the head'. Traditional cross-cultural research was insensitive to the cultural nature of theories, to the research methods, and to the tasks used by researchers. In order to avoid this pitfall it was necessary to understand the local practices of the groups studied.

From the late 1970's and throughout the 1980's, a common use of the notion of cultural practice in studies of mathematical cognition was associated with professional activities, that is, groups of people engaged in a specific craft or professional activity. Examples included Lave's studies of tailors in West Africa (1977), Nunes, Schliemann and Carraher's (1993) studies with street vendors, carpenters, fisherman, etc. in the North East of Brazil, Scribner's studies in a commercial dairy in the USA (1984). All these studies involved an ethnographic approach. Cultural differences between forms of mathematical thinking required in school and the forms of mathematics practised outside school were identified. Links between mathematical thinking, the cultural tools available and the context of the practice were also demonstrated. Though currently these links may seem obvious, one can see from David Carraher and Analucia Schliemann's (in press) recounting of the research with street vendors in Brazil that was not the case in the early 1980's.

Empirical findings from studies focused on cultural practices enabled the testing of Vygotsky's ideas in Western cultures and the development of a cultural psychology that has at its centre the notion of sociocultural mediation.

2.2. The Concept of Mediation in Cole's Cultural Psychology

Cole's re-thinking of the way culture and mind are mutually constituted culminated in the publication of 'Cultural Psychology: A Once and Future Discipline' in 1996. Though he adopted a 'cultural-historical' approach associated with the work of the Russian scholars Vygotsky, Luria and Leontiev, he suggested that the following are the main characteristics unifying the discipline:

- It emphasises mediated action in a context.
- It insists on the importance of the 'genetic method' [developmental analysis][1] understood broadly to include historical, ontogenetic, and micro-genetic levels of analysis.
- It seeks to ground its analysis in everyday life events.
- It assumes that mind emerges in the joint mediated activity of people. Mind, then, is in an important sense, 'co-constructed' and distributed.
- It assumes that individuals are active agents in their own development but do not act in settings entirely of their own choosing.
- It rejects cause-effect, stimulus response, explanatory science in favor of a science that emphasises the emergent nature of mind in activity and that acknowledges a central role for interpretation in its explanatory framework.
- It draws upon methodologies from humanities as well as from the social and biological sciences.

(Cole, 1996, p. 104)

The notion of **mediation** is the key to explaining the interplay between personal development and cultural practices. The concept of mediation as used in Cole's cultural psychology has its origins in Vygotsky's writings. It emerged from a distinction between psychological functions that assumed a direct access to the world (e.g. a young baby's perception) and functions based in indirect (or mediated) access to the world. For instance, Vygotsky (1978) referred to the use of knots as a tool in the function of mediated memory. In his view mediating devices, such as knots, numeracy systems and tools, extend the function of the mind beyond its biological dimension.

Most cultural psychologists would agree with the basic principles underlying the characteristics listed above, such as that the study of human development needs to take a social-historical and social-constructionist perspective, but they diverge in identifying the key features of the mediation. Cole acknowledges this divergence and explains it in terms of conceptions of culture. His way of making culture a relevant concept to understand cognitive development was to think of it 'as a medium constituted of historically cumulated artifacts which are organised to accomplish human growth' (Cole, 1995, p. 35).

Though common sense makes one think of an artifact as a material object, Cole described it as both material and conceptual. For him 'an artifact is an aspect of the material world that has been modified over the history of its incorporation into goal-directed human action. By virtue of the changes wrought in the process of their creation and use, artifacts are simultaneously *ideal* (conceptual) and *material*.' (Cole, 1996, p. 117). Very often the expression 'cultural tool' is used with the same meaning as cultural artifact. Vygotsky (1978) used the word 'tool' in connection

---

[1] Vygotsky used the term *genetic* to emphasise that the essence of psychological functioning could only be captured by methods and forms of analysis that study the origins and development of these phenomena.

with a physical artifact and reserved the expression 'sign system' to refer to mental artifacts. In this chapter the word tool is used to refer to both mental and physical artifacts. In the context of mathematical practices a tool can refer to a ruler, to a calculator, or to a system of signs, such as the base ten counting system.

The version of cultural psychology predominant in the studies of development of mathematical thinking and cognition has been in line with that of Cole. Studies on the uses of mathematics in out-of-school practices strongly supported the principle that any function of the mind is 'artifact/ tool' mediated. In addition, these studies were used to support the notion that mathematical cognition is situated in the context of the practice (Lave, 1988). Psychological differences, such as in learning or performance in mathematical tasks, were explained in terms of differences in the sociocultural practice. But, these explanations were not sufficient to clarify why the same group or the same person can be successful in one practice yet can experience difficulties in another. Why did the Brazilian street vendors learn mathematics in the streets, but have difficulties in learning school mathematics? Why did some of them do well in both practices while others did not? Why did children like Alberto (see Civil and Andrade's chapter) successfully acquire mathematical skills related to out-of-school practices but struggle with school mathematics?

In short, Cole's version of cultural psychology has as a strength an emphasis on tool mediated action rather than pure cognition. Emphasis on action enabled the recognition of the heterogeneity of psychological processes of groups engaged in distinct social and cultural practices. It also allowed recognition of the distributed nature of human cognition. Thus, the action of a person does not need to be situated in one mind, but rather in an activity system. However a weakness of this type of approach is that it does not yet provide a satisfactory account of within-group diversity.

2.3. *Some Criticisms of the Concept of Mediation in Vygotskian-Based Research*

Retrospective analyses of research in the West influenced by Vygotsky's ideas show that the studies on cultural mediation tended to focus on the relationships between the specific properties of the tool as a sign system and cognitive functioning (Abreu, 2000; Forman, Minick, & Stone, 1993). The studies that explored social mediation also had a very defined focus. The main objective was to illustrate the process through which immediate social interactions with more capable peers contributes to the learner's appropriation of cultural tools. Sometimes this type of analysis was carried to extremes, which led to criticisms that some descriptions of learning in the Zone of Proximal Development were nearly behaviourist.

'The zone of proximal development addresses how the child can alter his behavior by adopting my behaviour to become more like me. (...) [I]ts use can come perilously close to a description of learning as neobehaviorist shaping of behavior. It is especially true when the adult's role is described as a series of carefully arranged steps and teaching skills (e.g., 'raise the ante,' 'communicative ratchet,' 'extension') and when the child's contribution as *tabula rasa* is to absorb the language and structure from the adult input.' (Litowitz, 1993, p. 190)

Indeed this very peculiar focus is thought to have been a problem in developing an approach that explains the emergence of differences between individuals who in one way or another have connections to specific communities of practice, such as, children from similar ethnic backgrounds. Firstly, the studies that focused on the mediating role of the tools were quite often conducted within one practice. They tended to focus on individuals already engaged in that practice. Thus they did not provide any accounts of those who dropped out or who failed to achieve the required levels of mastery of the skills. This has been specifically pointed out in terms of a need to include a developmental dimension in the studies (Abreu, 1993; Saxe, 1991; Van Oers, 1998).

Secondly, the studies of interactions in the ZPD focused on situations of convergence, in which adults and children shared a view about the value of the skill to be learned. Again this resulted in 'descriptions of perfectly orchestrated dyads' (Litowitz, 1993, p. 187) so that success was the norm. Duveen (1997, 1998) has been arguing that behind this focus of research there is a particular conception of culture and social world. For Duveen: 'One limitation of Vygotsky's sociocultural psychology is that it portrays culture itself as a relatively homogeneous semiotic system; there is a seamlessness and continuity in its description which is especially evident in his discussion of internalisation (cf. Duveen, 1997). However, the social world is actually far more heterogeneous, it is a world structured around differences, around relations of power whose influence also structures the production and reproduction of knowledge' (1998, pp. 463–464).

These criticisms of the uses of the notion of social and cultural mediation are similar to the position Bishop (chapter 8) takes towards the 'educational assumption of consonance'. He also takes the view that assuming cultural consonance as tacit, accepted and unproblematic become untenable as soon as knowledge is viewed as socio-culturally based. Current societies are becoming more and more multicultural. The predominant conception of mediation – which emerged from the studies that treated learning primarily as a process of enculturation (see Cobb, 1995, and Bishop, chapter 8) – is not sufficient to understand encounters of experts and learners from different cultures. To get insights into this may require treating learning primarily as a process of acculturation or cultural production as argued by Bishop (chapter 8).

The next section attempts to illustrate how the focus on the experience of transitions brings a new dimension to the interpretation of cultural and social mediation.

3. TRANSITIONS BETWEEN PRACTICES: RE-VISITING THE NOTION OF MEDIATION

3.1. Initial Observations

I first became aware of the need to consider relationships between practices when doing research with school-children in a sugar-cane farming community in the North East of Brazil (Abreu, 1993). Prior to the study with the children I conducted ethnographic research with the farmers and described their specific mathematical practices. This farming mathematical knowledge was distinctive and historically prior to school mathematical knowledge. School was not compulsory when some of the farmers were children, thus a substantial number of them were unschooled and others went to school when adults.

For the school children the two forms of practice co-existed in the community. Attending school was compulsory, but engagement in farming varied. Consequently, some children learned some farming mathematics prior to learning school mathematics, while for others this order is inverted, and yet others never learned the farming mathematics although they lived in the community. This is an issue that has interested me: why if both sets of tools were equally important was there selective transmission and appropriation by new generations?

When I engaged the farmers in conversations, which involved relationships between farming and other local institutions, such as transactions with financial institutions or the schooling of their children, I noticed that, in these circumstances their tools stopped being seen as equally important. In this case, school tools were ranked as superior. The exposure to technological innovation and modern institutions (e.g. schools, banks) over time has raised farmers' awareness that some forms of knowledge were more powerful than others, and also that some were more acceptable than others. For instance, a contract in a bank could either be signed or stamped by a finger print. The farmer who signed might be functionally as illiterate as the one who stamps his finger print, neither of them being able to read the contract. However, the first method enabled the person to feel part of the literate society, and the second assigned the person to the category of 'illiterate'. The ability to sign was then highly valued by the group, while the use of the finger print was a reason to be ashamed. The same applied to the whole of traditional farming mathematics when compared with school mathematics. I referred to this phenomenon as a group's valorisation of knowledge. This seemed to reflect the status of the practices in the wider social structure.

What was clear at this point was the acknowledgement that to explain differences among the children the analyses had to go beyond cultural-tool mediation. Reflecting on the farmers' accounts I hypothesised that the perceived status of the cultural-tool in a social structure, or the social valorisation, was one of the dimensions of mediation that deserved to be further investigated. The next section explores the extent to which this double character of the mediation processes (cultural-tool mediation, and social-value laden mediation) was manifested in the empirical research presented in the previous chapters.

3.2. *Mediation in the Studies Reported in Chapters 2–6*

Cultural-tool mediation

Looking back at the empirical evidence reported in the previous chapters one can conclude that in agreement with the view of prominent cultural psychologists such as Cole (1996) and Bruner (1996), there is support for the notion that transitions between mathematical practices might expose students to distinct cultural tools. One can also see how a lack of awareness of this type of mediation can cause problems in learning situations. A compelling example of this was provided by Gorgorió and her colleagues when they recounted the story of Mohamed in the following terms:

> 'Mohamed, an immigrant student arrived into Catalonia and, when he was first at school, was diagnosed by his teacher as 'not knowing the basic mathematics algorithms'. He spent two years learning to add and beginning to subtract, not being able to communicate with the teacher. The following year, he moved to another town and his new teacher asked him if he already knew how to subtract when he came in. Mohamed, who already was able to understand and speak Catalan, was surprised at the question and showed his teacher 'his way' of doing 314 minus 182.'

Gorgorió et al's observation supports the key principle of cultural psychology that to understand human performance it is necessary to look at the interaction between the agent and the cultural tool. Wertsch's (1998) discussion of how someone will solve a multiplication problem can help us to clarify what happened to Mohamed. Suppose that a person was asked to multiply 343 by 822 and gave the answer 281946, and then that he or she was asked how the solution was obtained, Wertsch hypothesised that the person could reply: 'I just multiplied 343 by 822!' and if necessary produce a standard written algorithm as a demonstration. Wertsch used this example to question who solved the problem: the agent alone? Certainly not: he or she used a cultural tool – an algorithm for multiplication – to mediate the solution of the problem. Moreover he argues the role of the tool becomes even more salient if the person is asked to do a similar multiplication, without using the familiar algorithm. Suddenly the person's competence can be threatened. Some people could find it difficult to carry out the same operation without employing the particular tool they have mastered. Wertsch suggests that it seems more appropriate to say 'I and the cultural tool I employed' (1998, p. 29) solved the problem. Now looking back at what happened with Mohamed it seems clear that the initial assessment of what he was able to do was based on the agent acting alone.

Recognising the mediating role of tools, however, would be just a starting point. Several episodes in this book show that getting access to tools linked to practices in which one does not have direct participation is a complex process. To reflect upon this particular aspect Santos and Matos analysed situations where the *ardinas* were acting as 'informants'. This required re-framing their mathematical actions verbally to allow the researchers to learn about the process. Describing the case of Manitu they observed that when he did his calculations without verbalising he followed the expected sequence. However, he got confused when he was trying to verbalise what he was doing. Santos and Matos suggested that when Manitu was faced with the need to verbalise his thinking he was taken away from his *ardina* practice. He lost the secure points of support that assured the correct calculations, and he had to engage in a different practice, one in which he was not confident. Though this particular study was conducted within the *ardinas* practice and the major focus was on encompassing transitions, Santos and Matos speculated that when the *ardinas* assumed the role of informants they appeared to evoke a school frame. In evoking this particular frame they expressed difficulties in the use of particular tools, which in fact do not arise in the context of their *ardinas'* practice. Different needs for the communication of knowledge in the context of specific practices have to be addressed when exploring the mediating role of tools.

Tatheer Shamsi, the co-author of chapter 5, also recounted similar experiences when she observed Jafar – a year 2 British-Pakistani boy – working in his mathematics lessons. During her visits to his mathematics lessons she sat at his table and let him do his work and interact with the other three colleagues. She intervened only if asked, or when he completed each task. At this stage she would then ask Jafar to explain to her how he solved the particular task. What Jafar told her about how he was doing his sum did not always match with her observations of how he was doing it. On the day Tatheer had followed him in a lesson on subtracting with two digit-numbers, which required borrowing from the tens, she wrote in her diary:

> 'I feel that he sometimes justifies his strategies by voicing what is expected, where he may actually be using alternative strategies (mental arithmetic), but does not deem them acceptable, so he introduces other strategies when explaining'.

Later on Jafar confirmed to Tatheer that he was doing it in his head, and then demonstrated by counting on his fingers. So, it seems that he was evoking a frame related to the way his mother used to support him at home in order to explain to the researcher how he was solving the subtractions at school. He seemed to be moving back and forth between the requirements of the new strategy (as requested by the teacher) and his previous way of doing subtractions with which he felt confident. Both examples showed that knowing how to do something (using a tool) is different from explaining how one did it. In addition they also point out the need for more research both on how people communicate their thinking and on cognitive tensions which may arise when one is required to use different cultural-tools.

As illustrated in chapters 5 and 6 the difficulties in understanding a practice in which one does not have direct participation were also experienced by parents. Like the researchers, very often they were dependent on the children's or informant's explanations about how they used particular cultural tools. The children's school mathematical practices differed from their parents' own experiences. For some parents it varied because of educational reforms. Thus, the way parents learned mathematics when young was different from the way their children are taught. For other parents the differences were accentuated due to geographical movements (immigrant parents who went to school in another country and learned in a different language from their children). Even when a parent succeeded in figuring out the differences between tools, such as between the procedure they use to carry out a subtraction and the procedure followed in a child's school, this did not mean that he or she had acquired the skills to bridge the two procedures.

Social value-laden mediation

Gorgorió and her colleagues also demonstrated that the teacher's sensitivity to cultural mediation is only one part of the equation. This was illustrated by the way students reacted to the activities set up by a teacher who was attempting to help them to draw on tools and contents with which they were familiar. As a homework task the teacher asked the students 'to think, for tomorrow, of a mathematical problem or situation that can be linked with this photograph (of a rural market with a woman selling)'. The next day Miguel, a 16 years old gypsy student, who works cleaning houses, and who helps his family by selling in the weekend street markets, interpreted the task in the following terms: 'This was a trick! There is no mathematics problem, the woman has never been to school, she does not know mathematics.' In a follow-up interview Miguel expressed a very low opinion of his group regarding mathematical knowledge, and he was sure that if his people knew mathematics they would not be selling in street markets. This episode suggests that (i) Miguel interpreted the task with reference to what counts as legitimate mathematical practices and how they position people in the wider social structure; (ii) for this task to serve as a bridge enabling Miguel to bring to the classroom mathematical skills and understandings from outside school practices he would have to negotiate the meaning of what constitutes legitimate mathematical practices.

Gorgorió et al's observations are a replication of findings obtained in previous studies with school-children in other countries. Investigations with Portuguese children (Abreu, Bishop, & Pompeu, 1997), Brazilian children (Abreu, 1995b), and British children from Anglo and Asian backgrounds (Abreu, Cline, & Shamsi, 1999) all indicated that they developed ways of categorising some practices of their community as involving mathematics and others as less mathematical. Secondly, the findings concerning tasks that required them to choose the best and worst pupil in school mathematics suggested that children also associated performances in school mathematics with given social identities. For instance, more adults in a white-collar profession (office workers) were chosen as the best in school mathematics and conversely those in a blue-collar profession and other low social status practices as the worst. Since most of these studies were conducted with children from non-mainstream groups it was also the case that very often the children denied the existence of, or devalued, the mathematical tools associated with their home or out-of-school practices. I called this process the **valorisation** of mathematical practices, and have been arguing that to understand how people experience the relationship between their practices this needs to be treated as a key feature of the mediation.

Bishop in chapter 3 extends this process of categorisation to distinct school mathematical practices. He gave two examples illustrating how pupils from immigrant families from Eastern Europe perceived the school mathematical practices of the Australian school as less demanding. One of them, Dan, a Year 9 student from Eastern Europe, Georgia, described the organisation of his school mathematical practices in Georgia as different from that in Australia. He mentioned that in Georgia 'they wouldn't let you talk in class, you would be sitting down by yourself and working, working, as we have a lot more work in Georgia than this, a lot more'. When the interviewer asked Dan 'So you have the experience of two different systems; what do you think you prefer?', he replied 'Well I think that the Georgian system is better in terms that you would learn more.' Bishop also illustrated how perceiving one type of practice as superior to the other influences the way parents support their children's transitions between the school mathematics of their home country and the one practised in the host country. Thus, for instance, Jam, a Year 9 student from Russia, explained that though he is in level 4 at school his father only rates him in level 3 'because when we came here, like our level went down, like didn't do anything in primary school and so on. And when father compares it to Russia he would put me as a 3.'

It would not be difficult to find more examples in the empirical chapters to support this double character of the mediation process. As predicted by those authors that criticised the neutral-free approach to social mediation, the value-laden character of the process was heightened in transitions, which required adaptation and change. At this point, I would like to continue outlining a psychological perspective which emphasises the mediating role of cultural tools, but which also examines the mediating role of social valorisation.

4. INCORPORATING VALORISATION INTO A CULTURAL PSYCHOLOGY PERSPECTIVE ON MATHEMATICS LEARNING

4.1. Valorisation as a Feature of Social Practices: The Social Plane

The idea that to understand the impact of a cultural tool in cognition it is necessary to address links between the specific cultural practice where the tool is used and the broader social system has also been addressed by Wertsch (1991). He introduced the notion of 'privileging' to explain how groups and individuals come to make selective uses of tools. For him 'Privileging refers to the fact that one mediational means, such as a social language, is viewed as being more appropriate or efficacious than others in a particular sociocultural setting. (...) It is concerned with the fact that certain mediational means strike their users as being appropriate or even as the only possible alternative, when others are, in principle, imaginable' (Wertsch, 1991, p. 124). Expanding his argument he recently argued that '.... given that sociocultural settings inherently involve power and authority, any analysis that focuses on cognitive-instrumental rationality alone would have to be viewed as having essential shortcomings'(Wertsch, 1998, p. 64).

As exemplified in chapter 5 one strategy to incorporate the notion of valorisation into a model that also considers tool mediation is to think about the mathematical knowledge of a group in terms of social representations. From a social representations view each mathematical tool can be seen as a coin. One side of the coin relates to the technological characteristic of the tool (e.g. base five counting system). The other side relates to the value social groups associate with the tool, or its valorisation (e.g. base five counting systems are perceived as belonging to low status social groups) (Abreu, 1995a). As Moscovici (1998) remarked 'representations, like money, are formed with the double aim of acting and evaluating' (p. 244). With this notion the cultural nature of the mathematical tools is preserved: they continue to be seen as products of a cultural heritage. But, in addition they stop being seen as existing in a social-vacuum to become owned at some stage by social groups that belong to a social order.

From this perspective cultural practices become associated with particular social groups, which occupy certain positions in the structure of society. Groups can be seen as mainstream or as marginalised. In a similar vein individuals who participate in the practices will be given, or come to construct, identities associated with certain positions in these groups. The social representation enables the individual and social group to have access to a 'social code' that establishes relations between practices and social identities.

Gorgorió, Planas and Vilella in their chapter gave several examples illustrating how the experiences of the learners in their mathematics classrooms were filtered through representations they held about what counts as 'good school mathematical practices'. Their examples illustrate the links between valorisation and social identities. Thus, for Saima the gender identity was the main mediator of what she valued as proper mathematics. She could not see any value in doing the same mathematics as boys, since she did not want to have a boy's job. For Miguel the mediator was the status of the job in society. He did not see any value in discussing the uses of mathematics by groups that he saw as socially marginalised.

Other examples included students' representations of types of behaviours valued by their teachers. As illustrated by Bishop (chapter 3) Gor's understanding of the act of asking for help is not simply negotiating the understanding of a particular mathematical tool, but rather negotiating his level of competence (which can be taken as an indicator of a social identity as mathematics student). While some of the representations children expressed to the researchers might have been constructed in the current schooling practices in which the children were engaged, others seem to have been acquired in their home backgrounds or previous schooling in other countries. Thus, the social dynamics that shaped the immediate relationships with the teachers and researchers cannot be understood solely in terms of the 'here-and-now'. They only gained visibility when attention was given to the fact that as a social actor the learner participates in multiple practices.

Though social representation theory helps to locate practices in the social structure and to show how they relate to social identities as 'givens', it is still necessary to explain how these are re-constructed at the personal level.

4.2. Valorisation as Re-constructed by Learners: The Personal Plane

An Analogy Between Cultural and Social Development

From a cultural-tool mediation perspective learning has been described in terms of Vygotsky's 'general genetic law of cultural development', which asserts that:

> 'Any function in the child's cultural development appears twice, or on two planes. First it appears on the social plane, and then on the psychological plane. First it appears between people as an interpsychological category, and then within the child as an intrapsychological category. (Vygotsky, 1981, p. 163 cited in Wertsch, 1985 p. 11)

According to this law cultural tools are part of the cultural heritage of specific groups. They have an independent existence that can pre-exist that of individual participants of the practice. They may or may not be shared by all the participants of a practice as documented in studies of distributed cognition. And, as a resource that belongs to a particular social group they can be transformed. In order to become psychological the cultural tool has to be re-constructed by the individual.

This interaction between the 'given structure and resources' of a social practice and the psychological re-construction by individual participants is thoroughly examined in Santos and Matos chapter. As the authors remarked the arena in which the ardinas' practice of selling newspapers is located is 'a pre-existent entity independent from the participants [and] has its own evolution according to forces and relations that emerge from its nature (an economical institution) and position within the society as a whole'. They also provided clear examples of how the *ardinas* come to use specific cultural mediators to support the way they do their calculations. Thus, for instance, they observed that 'the sequences of mental calculation were structured with supporting elements that concerned values which were strongly related to the elements of the monetary system and to the price of the newspaper'.

It would be too simplistic to reduce the detailed analysis of the *ardinas'* practices in terms of re-construction of the mediating cultural tools. As Santos and Matos showed, the practice was socially represented 'as a work for young boys and not as a profession for adults', and this framed participation for short periods of time. This process of the dominant social representation of the practice impacting on forms of participation of individuals seems to suggest that value-mediation can also be explored first on the social plane and then on the psychological plane. However, they also discussed how the ardinas as individuals had different motives, which shaped the way they engaged and constructed their participation in the practice. This observation suggested that they were not passive reproducers of a social order, but exerted some sense of agency in their trajectories within the *ardinas'* practice. Next an approach that incorporates analysis at these two levels, without reverting to an individualistic perspective, is outlined.

Valorisation and social identity

Another retrospective look at data presented in various chapters will show that at a very young age children were already ranking some types of mathematical practices as being more valued than others. Thus, for instance, Civil and Andrade quote an eight year-old saying that:

> 'I hate math because I don't like low numbers, I like high numbers. They are *harder* and you get *smarter.* With low numbers, it's too easy'.

In chapter five (Abreu, Cline & Shamsi) there is also evidence that a child as young as seven years attributed more importance to the teacher's than to his mother's mathematics. So, one can claim there is evidence supporting the notion that children develop psychological understanding of the social valorisation of the practices they participate in. In addition chapter 5 illustrated that children were able to articulate and justify their preferences for one practice in relation to the other. These preferences could be based on a cognitive dimension of the process of learning, but they were also justified with reference to affective and social comparison dimensions.

As briefly outlined in chapter 5 one way of incorporating the above elements into a theory was to see mathematics learning as involving construction of social identities. Drawing on Tajfel's (1978) writing, social identity can be defined as involving two clear components: membership and position. Tajfel asserted that the concept of social identity is limited to 'that part of an individual's self-concept which derives from his knowledge of his membership of a social group (or groups) together with the value and emotional significance attached to that membership' (Tajfel, 1978, p. 63). Differentiating between membership and position has helped to operationalise the concept in our empirical studies. Membership defined belonging to a group, e.g. home group by birth, and position related to particular ways of re-constructing that membership. Position reflects both social understanding, knowledge of the location of a practice in a social structure and personal agency. This distinction was noted by Lloyd and Duveen (1992) in their studies of how children develop their gender identities. They demonstrated that children who shared knowledge of social representations of gender in their society, still developed distinct individual positions, reflecting degrees of acceptance or resistance to specific given social identities. The concept of position enabled us to account for the fact that children from the same home group (e.g. ethnic group) participated differently in the group's practices.

To sum up, valorisation of practices when re-constructed at psychological level gives the individual access to social identities. In the process of transition between practices these identities can be experienced in consonance or in conflict. In situations of conflict, it is argued in this book, it will become obvious that the construction of the mathematical knowledge is subordinated to identity construction (see Presmeg, chapter 9). In order to explore this issue further the concepts of identity chaining, competing identities and projected identities will be introduced.

Identity chaining

Valsiner (2000) uses the concept of identity chaining to explain how identities associated with specific cultural practices can be hierarchically organised by social institutions. The consequence for the individual can be a process of identity fusion. For instance, school as an institution can promote a form of identity where 'I as a Student' is suggested to be superior to 'I as I', and also includes the latter. Consequently, the 'I as I', or the emerging personal identity of a child in the home context can be fused into the identity promoted by school practices.

Among the consequences of school practices that promote the above pattern of identity chaining Valsiner stressed: (i) Children's rejection of parents' definitions of proper behaviour; (ii) Emergence of identity conflicts in the student; and (iii) The institutional appropriation of the family by the formal educational system. Though it would not be difficult to find examples to support Valsiner's ideas I would like to suggest that 'identity chaining' is not unidirectional. Throughout the empirical chapters there are various examples of students that seemed to be rejecting the 'I as a Student' identity given by the school in favour of the 'I as member of a specific home group'. This was illustrated, for instance, by the case of Alberto (Civil & Andrade's chapter) and by the case of Saima (Gorgorió's et. al. chapter).

Competing identities

One hypothesis explaining the multiple directionality of the chaining of identities is that of a progressive construction, which may be linked to the development of a sense of competence in practices that can be experienced as competing. For instance, in my previous studies with school-children in Brazil and Portugal they were often engaged in out-of-school activities that relied on 'informal' mathematics. Thus several of those who were failing in school mathematics were competent in other forms of mathematics related to their jobs. The job they did might have low status in the wider social system, but it still provided them with membership in a group in which their sense of their own competence was a source of personal pride. In this context it was the out-of-school activity that afforded the child a positive social identity. This dynamic may explain the desire and attempts of some of the children to drop out of school.

Looking at competing identities from Beach's (1999) perspective introduced in chapter one is also illuminating. He argues that when studying transitions between social practices it is necessary to take into account the direction of the movement. In his view lateral transitions, which involve a single direction (e.g. school to work) have different consequences than collateral transitions, which involve back and forth direction (e.g. home to school). He proposed that the direction of the movement creates distinct couplings. The term coupling is derived from biology and refers to coevolution as a changing relationship between different systems over time. In taking the developmental coupling as a unit of analysis for studies of transitions Beach assumes that it encompasses aspects of both changing individuals and changing practices and that individuals move across space time and changing social activities. Using this notion he showed that transitions between school and work mathematical practices were differently constructed by students (lateral transition) and by shopkeepers (collateral transition) in rural Nepal. For the students the transition was lateral because their participation in school preceded and was replaced by moving to work. For the shopkeepers the transition was collateral because the two practices were simultaneous. Schooling was not available in the village when the shopkeepers were at school age. As adult shopkeepers they were voluntarily attending school (adult education classes) to gain access to written literacy and arithmetic.

Beach found that the direction of the transition affected the coupling between the individuals and the practices. Both students and shopkeepers developed new strategies that did not belong either to 'schooling' or 'shopkeeping', but emerged as a product of the transition between the two. However, the new strategies were different for students and shopkeepers. The students showed resistance in moving away from written notation associated with their schooling, while the shopkeepers easily disregarded the written notations not meaningful in their practices (e.g. the arithmetic operations signs). For Beach a fundamental reason why the direction of the movement affected the transition was linked to the power and status of the practices in which one was or had previously been engaged. Thus, though the students were becoming shopkeepers, before that they spent ten years in school learning a form of mathematics that in Nepal society was perceived as 'superior'. On the other hand, the motivation for shopkeepers to go to school was not to become a student, but to access forms of knowledge that could enhance their economic situation.

From Beach's perspective the case of Alberto (Civil & Andrade, chapter 6) and the case of Kashif (Abreu, Cline & Shamsi, chapter 5) are not just two cases of children exposed to competing identities at home and at school. Though they both refer to children from immigrant families and who were having difficulties with school mathematics, the type of transitions they were involved in were quite distinct. Alberto as a first generation immigrant was exposed to a lateral transition. His sense of competence was associated with life in his home country, and this seemed to be a source of resistance to adaptation to school. For Kashif, a second generation immigrant, the source of competing identities was primarily associated with collateral transitions. His rejection or undervaluing of home mathematics in relation to school mathematics derived from his mother's involvement in helping him with his school work.

Projected identities

Another hypothesis about the possible direction of the chaining, and one which was explored in chapter 5, is the role that parents play in projecting social identities for their children, and therefore in intentionally encouraging them in certain practices and not in others. Though there is very little research in this area specifically related to mathematics, the idea is not totally new in the field of social and cultural psychology. In 1995 Cole wrote: 'Of crucial importance to understanding the contribution of culture in constituting development is the fact that the parents' projection of their children's future becomes a fundamentally important material/cultural constraint organising the child's life experiences in the present, because, as copious research has demonstrated, even adults totally ignorant of the real gender of a new-born will treat it quite differently depending upon its symbolic/ cultural 'gender'' (Cole, 1995, p. 37). In the gender research the symbolic / cultural gender has been experimentally manipulated, for instance, by dressing and giving names stereotypically as either boys or girls (e.g. blue versus pink clothes). The findings of this research showed that adults confronted with the babies behaved in culturally marked ways depending on the symbolic gender markers. Duveen (in press) recently elaborated this line of thinking when introducing the notion of 'extended identity'. By this he meant that in the process of child development very often how one is identified by others precedes one's own self-identification. Taking as an example the case of development of gender identity he suggested that 'before they are capable of independent activity in the field of gender (or any other social field) children are the objects of representations of other'. In the course of development they gradually internalise these identities and take positions that enable them to function as competent members of their community.

Again the initial observations for this type of explanation in relation to mathematical practices originated in my studies with Brazilian and Portuguese school-children (Abreu, 1995b; Abreu et al., 1997). A careful analysis of children's accounts of what they were accustomed to do after school revealed that what could be described superficially as the same activity posed different demands for them regarding their engagement in home mathematics. These different demands resulted from the ways in which parents structured the activity for the children. Concrete examples analysed in Portugal and Brazil covered the case of girls shopping for their mothers. In both studies the parents of the high achievers remained in charge of the economic dimension of the shopping activity, so that the girls did not report mathematical goals. The opposite trend was observed for the low achievers. These children were required to take responsibility for the economic dimension of shopping, and reported the mathematical goals that emerged in the situation such as being aware of unit prices of products, best buy prices, etc. It follows that children in this situation were more likely to develop mathematical knowledge that could conflict with school mathematics. (It is relevant to note that in these studies 'high achievers' and 'low achievers' are school designations. These were based on teachers' assessments of the child's performance in school mathematics.)

When looking at practices through the valorisation lens it becomes clear that under the same knowledge constraints the practices could be differently organised, and that this can be a form of extending specific social identities. This was observed in parents' accounts of the role of home language in the support they provide to their children's school mathematics learning. For instance, chapter 5 showed examples of two parents who have similar access to mathematics in their first language (Urdu), or the same knowledge constraints, but who followed different directions in supporting their children. The same reasoning can be applied when analysing the practices of teachers and institutions. For instance, the BRIDGE project reported by Civil and Andrade can be viewed as an attempt to develop school mathematical practices that will allow children to respect and feel proud of their immigrant identities and at the same time to succeed in the host country.

5. SOME CONCLUSIONS

To conclude I would like to suggest that the focus on transitions represents a third wave in the way relationships between learning and cultural practices are addressed. Key characteristics of the studies described in this book have included:

- A notion of cultural practice that pays more attention to the broader institutional and macro-interactional context.
- A notion of cultural practices that pays more attention to the social valorisation of these practices.

Extending the notion of cultural practice along the above lines has enabled the authors firstly, to view a variety of 'school mathematical practices'. School mathematical activities that took place in the context of the family were distinct from those taking place in the context of the child's school. The reasons why the practices were distinct varied. Thus, they could be distinct because of the use of distinct cultural tools, such as that they were conducted in a different language or because the parents used a different approach to teach a mathematical concept. But, they were also different because of value oriented expectations. This notion of practice is in line with Miller and Goodnow's (1995) proposal that 'practices are actions that are repeated, shared with others in a social group, and invested with normative expectations and with meanings or significances that go beyond the immediate goals of the action.' (p. 7) and that 'cultural practices are not neutral; they come packaged with values about what is natural, mature, morally right, or aesthetically pleasing.' (p. 6).

Secondly, this conceptualisation of social practices enabled the researchers to stress and demonstrate the value-laden character of mediational processes. Gradually as the impact of value was explored the cultural psychology approach outlined in the beginning of this chapter was elaborated to incorporate both cultural-tool and value-laden mediation. The impact of the value-oriented nature of the practices in the experience of transitions is a central aspect in the 'symbolic approaches' to cultural psychology (Shweder, 1990; Valsiner, 2000). This aspect is also elaborated in the anthropological perspective developed by Bishop and in the sociological perspective of Bourdieu developed by Presmeg in the next two chapters respectively. The concept of valorisation is concerned with 'symbolic capital' (Bourdieu, 1995), with the prestige associated with the ways of knowing of specific communities of practice. It is also concerned with acculturating dimensions of mathematical practices, which are invisible, or not examined, when the only focus of learning is the psychological re-construction of cultural tool or skills.

Finally, a revised notion of the key aspects of the practice that mediate between the individual and the sociocultural practices also required a different view of the psychological experience. In this chapter it was argued that this experience can be described in terms of construction of social identities. Participation in practices enables the person to master cultural tools and to understand how these are socially valued. This also enables the person to be positioned. It is the positioning that places the person in the social space. He or she accepts, rejects and re-constructs identities. This however, is a process that unfolds through participation in the practices. Educators can create social contexts that force individuals to exclude certain identities (or to keep them apart, hidden). Parents project identities for their children and structure environments for their children development according to their representations (see the chapter by Abreu et al.). This social structuring alone, however, did not explain the emergence of the unique individual. But it gave us clues about certain patterns of transitions that are more likely to be conducive to success or social inclusion.

REFERENCES

Abreu, G. de (1993). *The relationship between home and school mathematics in a farming community in rural Brazil.* Unpublished Doctoral Dissertation, Cambridge, Cambridge.

Abreu, G. de (1995a). A teoria das representações sociais e a cognição matemática [The social representations theory and the mathematical cognition]. *Quadrante, 4*(1), 25–41.

Abreu, G. de (1995b). Understanding how children experience the relationship between home and school mathematics. *Mind, Culture and Activity, 2*(2), 119–142.

Abreu, G. de (2000). Relationships between macro and micro sociocultural contexts: implications for the study of interactions in the mathematics classroom. *Educational Studies in Mathematics, 41*(1), 1–29.

Abreu, G. de, Bishop, A., & Pompeu, G. (1997). What children and teachers count as mathematics. In T. Nunes & P. Bryant (Eds.), *Learning and teaching mathematics: an international perspective* (pp. 233–264). Hove, East Sussex: Psychology Press.

Abreu, G. de, Cline, T., & Shamsi, T. (1999). *Mathematics learning in multiethnic primary schools* (ESRC – R000 222 381). Department of Psychology, University of Luton.

Beach, K. (1999). Consequential transitions: a sociocultural expedition beyond transfer in education. *Review of Research in Education, 24*, 101–139.

Bourdieu, P. (1995). *Language and symbolic power*. Cambridge, Massachusetts: Harvard University Press.

Bruner, J. (1996). *The culture of education*. Cambridge, Mass.: Harvard University Press.

Carraher, D.W., & Schliemann, A.D. (in press). Is everyday mathematics truly relevant to mathematics education? In M. Brenner & J. Moschkovich (Eds.), Everyday and academic mathematics in the classroom. *Journal of Research in Mathematics Education Monograph.*

Cobb, P. (1995). Mathematical learning and small group interaction: four case studies. In P. Cobb & H. Bauersfeld (Eds.), *The emergence of mathematical meaning* (pp. 25–129). Hillsdale, New Jersey: Lawrence Erlbaum.

Cole, M. (1995). Culture and cognitive development: from cross-cultural research to creating systems of cultural mediation. *Culture & Psychology, 1*(25–54).

Cole, M. (1996). *Cultural psychology*. Cambridge, Mass.: The Belknap Press of Harvard University Press.

Cole, M. (1998). Can cultural psychology help us to think about diversity? *Mind, Culture and Activity, 5*(4), 291–304.

Duveen, G. (1997). Psychological development as a social process. In L. Smith, P. Tomlinson, & J. Dockerel (Eds.), *Piaget, Vygotsky and beyond*. London: Routledge.

Duveen, G. (1998). The psychosocial production: social representations and psychologic. *Culture & Psychology, 4*(4), 455–472.

Duveen, G. (in press). Representations, identities, resistance. In K. Deaux & G. Philogene (Eds.), *Social representations: introductions and explorations*. Oxford: Blackwell.

Forman, E.A., Minick, N., & Stone, C.A. (Eds.). (1993). *Contexts for learning*. Oxford: Oxford University Press.

Gay, J., & Cole, M. (1967). *The new mathematics and an old culture: a study of learning among the Kpelle of Liberia*. New York: Holt, Rinehart and Winston.

Lave, J. (1977). Cognitive consequences of traditional apprenticeship training in West Africa. *Anthropology and Education Quarterly, 8*(3), 177–180.

Lave, J. (1988). *Cognition in practice*. Cambridge: Cambridge University Press.

Litowitz, B.E. (1993). Deconstruction in the Zone of Proximal Development. In E. Forman, N. Minick, & C.A. Stone (Eds.), *Contexts for learning* (pp. 184–185). Oxford: Oxford University Press.

Lloyd, B., & Duveen, G. (1992). *Gender identities and education*. London: Harvester Wheatsheaf.

Miller, P.J., & Goodnow, J.J. (1995). Cultural practices: towards an integration of culture and development. In J.J. Goodnow, P.J. Miller, & F. Kessel (Eds.), *Cultural practices as contexts for development* (pp. 5–16). San Francisco, California: Jossey-Bass.

Moscovici, S. (1998). The history and actuality of social representations. In U. Flick (Ed.), *The Psychology of the Social* (pp. 209–247). Cambridge: CUP.

Nunes, T., Schliemann, A., & Carraher, D. (1993). *Street mathematics and school mathematics*. Cambridge: Cambridge University Press.

Ratner, C. (1999). Three approaches to cultural psychology: a critique. *Cultural Dynamics, 11*(1), 7–31.

Ratner, C. (2000). Outline of a coherent, comprehensive concept of culture. *Cross-Cultural Psychology Bulletin, 34*(1 & 2), 5–11.

Saxe, G. (1991). *Culture and cognitive development: studies in mathematical understanding*. Hillsdale, New Jersey: Lawrence Erlbaum.

Saxe, G.B. (1982). Culture and the development of numerical cognition: studies among the Oksapmin of Papua New Guinea. In C.G. Brainerd (Ed.), *Children's logical and mathematical cognition* (pp. 157–176). New York: Springer Verlag.

Scribner, S. (1984). Cognitive studies of work (Special issue: *The Quarterly Newsletter of the Laboratory of Human Cognition*, Vol. 6, pp. 1–50). San Diego: University of California.

Shweder, R.A. (1990). Cultural psychology – What is it? In J.W. Stiegler, R.A. Shweder, & G. Herdt (Eds.), *Cultural psychology* (pp. 1–43). Cambridge: Cambridge University Press.

Tajfel, H. (1978). Social categorisation, social identity and social comparison. In H. Tajfel (Ed.), *Differentiation between social groups: studies in social psychology of intergroup relations* (pp. 61–76). London: Academic Press.

Valsiner, J. (2000). *Culture and human development*. London: Sage.

Van Oers, B. (1998). The fallacy of decontextualisation. *Mind, Culture, and Activity, 5*(2), 135–142.

Vygotsky, L. (1978). *Mind in society: the development of higher psychological processes*. Cambridge, Mass: Harvard University Press.

Wertsch, J.V. (1985). Introduction. In J.V. Wertsch (Ed.), *Culture, communication and cognition* (pp. 1–18). Cambridge: Cambridge University Press.

Wertsch, J.V. (1991). *Voices of the mind*. Cambridge: Cambridge University Press.

Wertsch, J.V. (1998). *Mind as action*. Oxford: Oxford University Press.

CHAPTER 8

MATHEMATICAL ACCULTURATION, CULTURAL CONFLICTS, AND TRANSITION

ALAN BISHOP
Monash University

1 INTRODUCTION

Explanations and interpretations of under-achievement in mathematics have tended to move away from the cognitive domain in recent years. Mathematics educators have realised that the cognitive psychological program has been found wanting in terms of both explanations and implications for changing practice, bearing in mind that there is only so much that individual learners can do about their situation.

This general realisation has found action and voice in the research communities through analyses of anthropological and sociological constructions of mathematics education (Barton, 1996; Restivo, 1993). The related field of research on situated cognition (e.g, Kirshner & Whitson, 1997) is also relevant here, as it reminds us that, in becoming more knowledgeable about the social and cultural aspects of mathematics education, we should not forget that the teachers and the educational system are also dealing with individual learners. What is crucially important is the quality and the nature of the interaction between the learner and the social learning environment. This is particularly important in considering the nature of the experience of transition in the learning and the practice of mathematics.

In particular, this chapter will consider the role of ideas related to cultural conflicts and mathematical acculturation, in seeking to clarify issues, constraints, and potentialities in relation to transitions in mathematics learning. Thus the perspective taken will not focus on the individual learner, but on the acculturation process itself, its structures and instruments, and on the roles of those whom we might wish to consider as the 'acculturators' (Spindler, 1974). It will consider the learners through their experience of cultural conflicts, and will explore how the acculturation process interacts with learners' actions to co-construct the learning and the practice during the transition experience. Evidence will be used from previous chapters to explore these ideas in a variety of educational situations.

Acculturation is a construct borrowed from anthropology, and is contrasted with 'enculturation'. As Wolcott (1974) describes it succinctly this way: 'Anthropologists refer to the modification of one culture through continuous contact with another as

G. de Abreu, A.J. Bishop, and N.C. Presmeg (eds.), Transitions Between Contexts of Mathematical Practices, 193–212.

acculturation. Often one of the contact cultures is dominant, regardless of whether such dominance is intended' (p. 136). Enculturation by contrast is the induction, by the cultural group, of young people into their culture. The distinction for educational discourse hinges on whether one considers the educational encounter to be an enculturative or an acculturative experience. This chapter focuses on, and assumes, the second perspective.

By educational 'institution' in this chapter will be meant any formal or non-formal educational context for learning mathematics, using Coombs' (1985) descriptions. Though these contexts are very different in the eyes of the educational bodies concerned, for the purposes of this chapter, and this book, the issues are similar, with only the details of the institutions being different. For example, in Santos and Matos' chapter, one can consider that the *ardinas* had their own non-formal institution for learning the mathematics involved in newspaper buying and selling. It did not have a formalised organisational structure nor a formal physical structure, although the main square apparently functioned non-formally in that way. It did not have a formalised set of rules and written procedures, but as chapter 4 shows us there were plenty of non-formal, unwritten, rules and procedures for regulating the *ardinas'* mathematical practice. In that sense the institution of '*ardinas* practice' was as real and as powerful as any school in its influences on the transition experience.

In the next section of this chapter, the attention will be on the mathematical learners through their key transition experience of cultural conflicts. That section will explore the kinds of cultural conflicts experienced by learners in the previous chapters, and will consider an alternative way to view conflict. It focuses on the distinction between the ideas of conflict as essentially an aspect of difference and mismatch, or as part of a complementarity underlying explicit cultural interaction (Valsiner & Cairns, 1992, Shantz & Hartup, 1992).

In section three, there will be more interrogation of the idea of mathematical acculturation. In particular the concern will be with the various teachers of mathematics, their power and their role as 'acculturators' in the transition situation. In a similar way to the wide interpretation of 'institution' the idea of 'teacher' will be interpreted in the wider sense of anyone structurally involved in facilitating the learning of mathematics. Teachers in a generic sense are structurally part of every formal, and non-formal, educational institution and the power of the institution is manifested not only through its rules and values, but also through the ways the teachers interpret and mediate that power through the hidden and not-so-hidden curricula. The roles of parents and other adults will also be considered here together with the ways they relate to the teacher and the institution. However power as an interactional and 'productive' construct will offer a different perspective on the situation from the usual idea of power as 'sovereignty' (Popkewitz, 1999).

Section four will focus attention on how the site of mathematics learning can be socially reconceptualised in the transition process. There the role of the 'significant others', the learners' peers and classmates, will be explored, together with the significance of viewing the mathematical learning situation as a workplace with its own practices and 'borderland' discourses (Gee, 1992, and O'Connor, 1995).

The final section will discuss the relationships between the three key aspects of: the acculturation process and productive power, the borderland Discourse of the mathematics classroom, and explicit cultural interaction between teachers and learners. It will build on the idea of 'cultural production' as a viable mediational metaphor (see Levinson, Foley & Holland, 1996) not only for transition situations that provoke extreme cultural conflicts but for all mathematics educational situations.

2. CULTURAL CONFLICTS: FROM EXCLUSIVE TO INCLUSIVE PARTITIONING

In 'Mathematical enculturation: A cultural perspective on mathematics education', (Bishop, 1988) cultural conflict was not specifically addressed, although the ideas of cultural difference and similarity played a large role in the first part of the analysis, where different mathematical knowledge and cultural values were analysed. After demonstrating the pan-cultural nature of mathematical activity, however, the educational analysis in that book followed the enculturation path, in which the principal assumption made was of cultural consonance. In Bishop (1994) however, there was an elaboration of a research program on cultural conflicts, and this chapter continues that theme.

The assumption of cultural consonance was for many years tacit, accepted and unproblematic. However, as soon as mathematics became understood as culturally-based knowledge, largely through the work of the ethnomathematicians, the educational assumption of cultural consonance became untenable. It was then clear that many young people in the world were experiencing a dissonance between the cultural tradition outside the formal educational institution (for example in their home, in their part-time work practices, or in their previous country) and that represented inside the institution. This chapter however makes the more radical assumption that all mathematics education is a process of acculturation, and that every learner experiences cultural conflict in that process. However, cultural conflict need not be conceptualised exclusively in a negative way, as will be seen.

The construct of 'cultural conflict' grew out of educational research in the anthropological tradition. We can find it, for example, as a central idea in McDermott's (1974) classic chapter about 'pariah groups' whose children fail to succeed in mainstream schools. He builds on Barth's (1969) definition of pariah groups, who are those who are 'actively rejected by the host population'. According to McDermott, 'Students and teachers in a pariah-host population mix usually produce communicative breakdowns by simply performing routine and practical everyday activities in ways their sub-cultures define as normal and appropriate….The problem is neither 'dumb kids' nor 'racist teachers', but cultural conflict' (p. 173).

The idea that conflict is a necessary part of every educational encounter, is supported by Moscovici (1976) in his writings about social change. In his chapter 5 called 'The knot of change: conflict' he discusses the centrality of conflict in producing change. The book offers several provocative propositions, one of which is Proposition 3 which states: 'Influence processes are directly related to the production and resolution of conflict'(p. 98). Moscovici summarises his conclusions in this way: 'Let us repeat that conflict is a necessary condition of influence. It is the starting point and the means for changing others, of establishing new relations or consolidating old ones. Uncertainty and ambiguity are concepts and states derived from conflict' (p. 105).

Conflict, according to Moscovici, is thus a construct referring to affective states and situations, which are assumed to involve antagonists in some sense. However cognitive conflict is a well-known and well-researched explanatory construct that has been transformed by some into a pedagogical tool. To create cognitive conflict intentionally is accepted by many mathematics educators as a legitimate pedagogical strategy. However, working with the construct of 'cultural conflict' initially provokes a very different kind of response from teachers to its cognitive counterpart. In-service educational contexts, it is rare to find a single teacher who has admitted to *intentionally* putting their students in a cultural conflict situation. Among the reasons might be cultural sensitivity, or political correctness, but another reason reflects an awareness that as well as having a cognitive component, cultural conflict involves strong emotional and affective responses. Where cognitive conflict generates a cognitive response, the cultural conflict situation generates a cultural response, which is more than cognitive in character. Accepting this view, a corollary might be that cultural conflict situations should demand a sympathetic and understanding response from the teacher and peers. The teacher should be thinking less about creating conflict, as this will naturally occur, but much more about aspects of conflict resolution.

There has been plenty of evidence of conflicts in the previous chapters, which are clearly not cognitive conflicts in the usual sense. They are referring to the cultural nature of the learning situation, its norms, its values, the genre of its discourses, the non-verbal communication patterns etc. Whilst there is likely to be a cognitive component to these conflicts, we can see that there are also emotional and affective overtones indicating deeper and more fundamental aspects.

From chapter 2, the interview with the teacher Maria reveals part of the problem.

Int: Regarding working in small groups, how do they deal with it? Do they adapt to it? When they can choose to work in small groups or individually, which do they prefer?

M: (no doubt in her voice). Individually: immigrant students, when they arrive in our country, they have to face the fact that other students interpret their not knowing the language and the habits as being stupid. 'He does not understand us, we say white and he does black!' That is why, when working, if the students can choose their mates in the groups, nobody wants to work with the immigrant ones, because 'he does not know'.

Another aspect referred to by the same teacher is the following:

Int: Do the activities that the immigrant students develop outside school interfere with what they do in school?

M: There is a delicate moment, when they begin to go to the mosque. They have a fear of going there to learn: there, to fail, not to know is penalised, they are punished. When they begin going to the mosque, they become closed!... Boys are punished, it is not the same with the girls, probably because it is not so important whether a girl learns or not.

The issue of the lack of corporal punishment was also mentioned by the students in chapter 2:

Teacher: Sajid, are you not interested in what we are doing/
Sajid: Yes Miss I am
Teacher: Why do you not work then?
Sajid: In my country ... beat me, then I study. Here nobody beats ...

In chapter 6, Civil and Andrade discuss the case of Alberto who they say represents an extreme transition situation. Alberto didn't want to come to the USA, resented his family's move from Mexico to USA and didn't want to learn English.

In chapter 3, Bishop reports part of an interview with Tra, a recent immigrant student:

Int: Can you describe a situation in the maths class that was a bad time, a bad experience?
Tra: Not now but it was last year, I always do my work but there is no test and there isn't really test just assignments and they don't really show you what level you are. And I was so quiet as it was depressing as no-one noticed you...

These are just some of the examples of the cultural conflict situation in which immigrant students find themselves. But is this a productive way to consider the situation? Is it helpful for learners to know that they are experiencing cultural conflict? Is there anything that the teachers can do about the conflict situation of the learners? Does it have anything useful to say about teaching, learning and mathematical activity?

Valsiner and Cairns (1992) argue that there has been too much use in research made of the common-sense meaning of conflict, without sufficient exploration of other meanings. The common-sense meaning they consider to be an example of *exclusive partitioning*, whereby 'conflict can be defined as mismatch or difference, and assumes that conflict resolution becomes synonymous with a choice between different pre-existing preference options' (p. 24). This view essentially defines conflict as difference and assumes that by eliminating difference one resolves the conflict. In our terms the students either choose to adopt the host culture's norms or continue to be an outsider, it is an either/or choice.

In contrast, Valsiner and Cairns argue that *inclusive partitioning* preserves heterogeneity and enables conflict to be defined in terms of the nature of linkages between the differentiated parts of a whole. These linkages can be viewed as involving opposition between the parts. 'The opposites are inseparable as they are functional interdependent parts of the whole' (p. 25). The opposites, which include conflict as a sub-class of opposites, make it possible for the parts to co-exist. They give as an example the following: '... the 'harmonious play' of children may become transformed into a 'fight' and vice-versa, where both states of the children's relationships indicate 'friendship' by both children (consider an adolescent's claim that she 'fights with her brother because we are good friends')' (p. 25).

This inclusive partitioning approach can then enable us to see a wider picture. We earlier introduced into the language of this chapter the terms 'cultural consonance' and 'cultural dissonance' and now it is important to re-analyse them. In Valsiner and

Cairns' terms these two constructs could be considered as the opposite parts of a linkage concerned with 'cultural interaction'. The complementarity between them implies that each contains within it the seeds of the other. Dissonance contains within it aspects of consonance, since in order to disagree one must recognise some common features, and likewise consonance includes the raw material of dissonance. If these conditions are not met there is, by definition, no cultural interaction taking place.

Rather than seeking resolution of the cultural conflict by eliminating difference, one can imagine instead the possibility of engaging overtly and explicitly in cultural interaction, which will involve an alternating and reciprocating development of conflict and consensus, resulting continuously in both dissonance and consonance. Moscovici's proposition 3, which was quoted earlier in this chapter, states: 'Influence processes are directly related to the production and resolution of conflict', but it is not necessary for teachers to think that they need to intentionally create conflict. Conflict is only one part of the perception of cultural interaction. What the teacher should be doing is thinking about how to help to create the conditions for explicit cultural interaction to take place. Of course one implication of this is that it is possible, and arguably educationally desirable, for the learner to also help to establish the conditions for cultural interaction to take place. Whether conflict or consensus results from this cultural interaction will of course ultimately depend on the learner – it is the learner's response and construction of the whole cultural interaction situation that dictates what happens here.

In chapter 2 Gorgorió et al. present an example of this cultural interaction, potentially producing both conflict and consensus. The teacher had given the students worksheets for the session:

Nashoua: May I bring the books I had in Morocco?
Teacher: To show them to me?
Nashoua: No! To use them.
Teacher: What do you need them for?
Nashoua: To work with them, to know what the class is about!

Nashoua is clearly not happy about the situation but rather than just feel resentment she takes some initiative in the matter. The 'ball' is now in the teacher's 'court' and she must decide what her response will be, and whether or not she will engage in this explicit cultural interaction. It is not a normal situation for the teacher, and her response together with the continuing interaction will determine the extent of conflict and consensus that the student feels.

In chapter 4 Santos and Matos present another example where one can imagine that this cultural interaction will take place in the newspaper selling 'institution' of the *ardinas*. It concerns the use of the calculator:

> The difficulty he (Kode) showed in using the calculator (which is not used at school) shows how this instrument, despite being present in the ardinas' everyday life, is actually not part of the artifacts of their practice. We only observed the use of a calculator by some of the old-timers and even then it was just to calculate the value they had to pay. The calculator is an artifact of Disidori and Manu's practice when controlling the ardinas' payments, but it is not one of the artifacts of the ardinas' practice.

Although the calculator is not a part of their practice, one wonders how long it will be before it does become a part. It is already a part of their everyday life, and given some of the calculation complexities described in chapter 4, it could well be used more frequently.

Pursuing their argument further, Valsiner and Cairns (1992) add that: 'The crucial distinction within the class of conflict theories is between the (a) conflict maintenance and (b) conflict transformational notions.' According to the first theories, one seeks to resolve the conflict so that it no longer exists. In the case of mathematics education, that is not just an impossible, but also arguably an undesirable, goal. It ultimately rests on the view that cultural difference should be eliminated or at least ignored. The latter of course is certainly how cultural difference has often been viewed, commensurate with the view that mathematics education is a culturally neutral affair! According to the second class of theories, one treats conflict as always and necessarily present, but along with its complementary partner, consensus. The particular conflict will change, along with the source of the conflict, as development occurs, as cultural interaction continues, and as particular consensus is reached in part of the cultural interaction complex.

3. MATHEMATICAL ACCULTURATION: THE TEACHERS' ROLE AND PRODUCTIVE POWER

This section of the chapter focuses on the teacher as the prime agent of acculturation. Detailed research on the notion of 'home culture' has challenged the assumption of universal cultural consonance in mathematics directly (e.g., Abreu, Bishop & Pompeu, 1997) by showing that in many situations the home culture and the school mathematics culture can be conceptualised as being mutually exclusive. Moreover in Abreu's (1995) research context, the teachers appeared to want to keep it that way, made no reference to any out-of-school or home mathematical knowledge, had no interest in finding out anything about their students' out-of-school knowledge, and probably would not have been able to do anything with that knowledge even if they did know it. This is not only because of their own knowledge base, but also because of the formal institutionalised system of curriculum and examinations, and the social and professional pressure from their colleagues etc., which 'supports', sustains, and indeed controls their views. In the terms of the last section they were ensuring that no cultural interaction was made explicitly possible.

As was stated at the start of this chapter one key assumption made here is that a young person's mathematics education is necessarily an acculturation experience, with its accompanying emotional states and cultural consonances and conflicts that need to be understood, tolerated and arguably fostered. But in situations where there is extreme dissonance between the outside and inside institutional cultural norms, it is very unclear what the institution's, and the teacher's, acculturation strategies and indeed their educational task should be. The established theories of mathematics education, developed through a research history that has failed to recognise the essential cultural interaction at the heart of mathematics education, are at best misleading and at worst irrelevant and obstructive. The task of exploring the conceptual implications of mathematical acculturation seems to be of a totally different order.

If we consider education to be an intentional process, on the part of society and carried out by teachers, what then can we understand by 'intentional acculturation'? It certainly creates images of the worst kind of brain-washing and assimilative experiences which migrants to USA, Australia, UK, and elsewhere certainly suffered in the past. We can also sense the conflicts involved in Wolcott's (1974) chapter 'The teacher as an enemy' where he talks of 'antagonistic acculturation'. He was a 'white' person teaching Kwakiutl Indians in their village school in North America. He says of his experiences: 'I had never encountered teachers or pupils with whom I did not share relatively similar expectations regarding behaviors, values, and attitudes...At Blackfish Village my pupils and I shared few mutual expectations regarding our formal role relationship...If we were not at any one moment actually engaged in a classroom skirmish, it was only because we were recovering from a prior one or preparing for the next' (p. 144).

He uses the idea of 'teacher as enemy' not in the sense of entering into combat, although he says: 'on the worst days we may not have been far from it.' He uses it in the analogy of the prisoner-of-war camp, because as he says about that situation: 'While great hostility on the part of either group (inmates and captors) might be present in the relationship, it is not essential to it, because the enmity is not derived from individual or personal antagonism. Nonetheless, the captors, representing one cultural group, are not expected to convert the prisoners to their way of life, and the prisoners are not expected to acculturate the captors' (p. 145).

Wolcott continues his analogy by introducing teachers into the example: 'Suppose that along with the usual cadre of overseers the captors have also provided teachers charged with instructing the prisoners in the ways of, and particularly the merits of, their culture. The purpose of instruction is to recruit new members into their society by encouraging prisoners to defect, and achieving this by giving them the skills so that they can do so effectively' (p. 145). He argues that there are many instructive messages, and his paper continues to describe them, that emerge for teachers from considering this analogy in educational situations of extreme cultural mismatch. For example, he says this: 'Most important, the teacher realises the meaning that accepting his teaching may have for those prisoners who do accept it. It may mean selling out, defecting, turning traitor, ignoring the succorance and values and pressures of one's peers, one's family, one's own people.. .The teacher needs constantly to review what these costs mean to any human' (p. 147).

Making the transition from this image to the examples in the previous chapters, one can see this in the following example from chapter 2 as the teacher discovers that her attempt to 'encourage the student to defect' results in a sarcastic rejection of the idea:

Teacher: I see, you have been able to finish the activity.
Sheraz: (in a broken language) Very easy.
Teacher: (in a hurry, she is needed in another room) You will explain it to the whole group next day, will you?
Sheraz: I do not explain. Very easy.
Teacher: (slowing down) It is easy for you Sheraz, but may be, your mates or myself, we do not find it easy.

Sheraz: (with a disrespectful tone): What a teacher of mathematics you (are)! Better (stay) at home! Here students know more (than) teacher!

Clearly, in trying to clarify the nature of intentional acculturation, the role of this 'acculturator' is a crucially sensitive one, and is one that is related to their position and possession of power. Teachers are in a sense the agents of acculturation and are invested by the educational system and their institution with the necessary power to do this work. How is this power manifested in the transition situation and how does it relate to the cultural interaction discussed in the previous section?

Santos and Matos in their chapter vividly demonstrate this power that Disidori and Manu have. They are the experienced *ardinas* who also play the acculturator role in the *ardinas* institution:

> ...during payment, if any ardina faced a situation of disparity between what he expected to pay and what he was asked to pay, his behaviour was discrete and non-argumentative. He would try to delay payment, showing signs of doubt such as looking once again at the calculator screen or recounting the money. This way Disidori or Manu were forced to do their calculations again, paying more attention to the numbers turning up on the calculator. Some ardinas did not even hint at it, simply accepting that the value on the screen of the calculator was correct and handing over the money. At the end they would walk away with their heads low, feeling sad but without saying a word on the subject. If the observer approached them and talked to them about the matter they would often explain what had happened as proof that they had lost money during the sale, for instance, or given too much change back, or someone had stolen some newspapers, or else they had not received the correct number of newspapers from the start. Since Disidori's and Manu's calculation processes were completely transparent, they watched the numbers Disidori put in the calculator ('to see if he's doing the sum well') and assumed that the calculator never made mistakes and so the result was never defied.

In chapter 3 Bishop discussed the institutional and teacher practice of labeling students. It is used as a strategy for exerting power over the learners, as well of course for administrative purposes. Whether students are 'ESL', 'NESB', 'black', 'migrant', 'ethnic minority' or 'disadvantaged', being labelled in that way guarantees that one will not be perceived in a rich, multifarious, individualistic way. In chapter 2 we read in the interview with the teacher Maria:

Int: Do you have any materials adapted to immigrant students?

M.: I don't, not for them specifically. But I have some materials for less-able students and sometimes I use them for the immigrant ones.

Categories and labels are clearly very useful in performing various administrative functions, as was shown in chapter 3. They also permeate teachers' discourses, but the danger there is that the simplifying role of labeling offers little recognition of the variety of background experiences and abilities that the immigrant students can bring with them. Equating them in some way with 'less-able' students shows another negative aspect of the power relationship, even if it comes with a benign intention.

Becker (1962) has an interesting perspective on the powerful strategy of labeling, based on his research into deviance in society.

'The central fact of deviance is that it is created by society. I do not mean this in the way it is ordinarily understood, in which causes of deviance are located in the social situation of the deviant, or the social factors which prompted his action. I mean, rather, that social groups create deviants by making the rules whose infraction constitutes deviance, and by applying those rules to particular people and labeling them as outsiders. From this point of view, deviance is not a quality of the act a person commits, but rather a consequence of the application by others of rules and sanctions to an 'offender'. The document is one to whom the label has been successfully applied. Deviant behaviour is behaviour that people so label.' (pp. 8, 9)

Deviants are not therefore homogeneous. They share the label and the experiences of being so labelled and that is all. It is the same with immigrant students. They are not a homogeneous group, but they share the label and the experience of being so labelled, as does anyone who is perceived as an outsider. The teacher's power legitimises both her use of labeling, as well as her non-use of labeling, like this teacher from chapter 2, who responded to this issue by ignoring any cultural differences:

In the past I have been very worried about how to teach a multicultural group of students, but now I am clear about it. (...) We have to normalise the situation of the students that come from abroad. One has to treat all the students the same, without pointing out any differences. By pointing out the differences it is much easier to fall on discriminating them.

Presumably this teacher would have the same underlying philosophy as the other teacher in chapter 2 whose classroom looked like this:

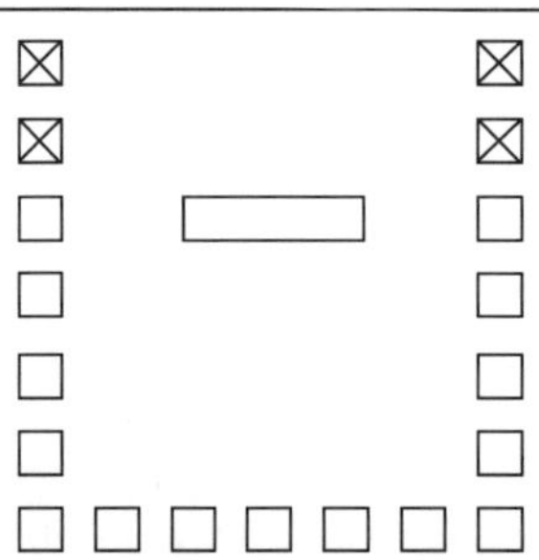

This is the schema of the tables' organisation in a mathematics classroom. It is a class where the teacher usually faces the group having the blackboard behind him. The teacher's table is the one in the centre and the ones crossed are where the immigrant students sit.

This strategy would certainly seem to avoid the labeling problem, but at what cost? How could the teacher recognise the richness of each individual without attending to their differences? The cultural and social norms of the classroom still have to be established, so will the host culture determine the norm and ignore the consequent cultural conflicts? The big question is how can differences be conceptualised and taken into account in a way that avoids discrimination?

These kinds of examples show us that the teachers face a complex responsibility with regard to their position of power in this transition process. They have many problems if they exert their power too strongly, through for example their use of simplistic labels, because all this does is to lead to their oppression of the learners and a denial of their unique qualities. Similarly they run the risk of creating other problems with their institutions and their colleagues if they don't use the labeling type of power strategies that are so familiar in educational institutions. This is once again where we need to reconsider the concept of power itself, much as was done with conflict in the previous section.

Popkewitz (1999) moves us on from Becker's ideas and offers us another way to conceptualise power that is more productive. He would argue that so far in this section the construct 'power' has been used as something that people own, which is the kind of power he calls 'sovereignty'. In his view: 'Power as sovereignty often creates a dichotomous world in which there are the oppressor and the oppressed, thus producing a dualism whose effect is to define particular social groups as unified entities …The dualism of oppressor/oppressed loses sight of the productive qualities of power, that is, how systems of knowledge generate principles through which action and participation are constructed' (p. 6).

Elaborating this distinction Popkewitz goes on: 'The productive elements of power move attention from identifying the controlling actors to identifying the systems of ideas that normalise and construct the rules through which intent and purpose are constructed in action…The study of the effects of power enables us to focus on the ways that individuals construct boundaries and possibilities' (p. 6). We can see that labeling is part of boundary construction, and it is interesting to note how similar this point is to that made by Becker earlier about 'deviance'. It is not the people who are deviant, nor who are powerless, in some homogeneous way, but it is the 'system of ideas and knowledge that normalise and construct the rules' in Popkewitz terms. It is that same system that labels and defines 'deviants', 'immigrants', and 'less-able students'.

In our context then, this construction of 'productive power' recognises that teachers do not need to just accept unthinkingly the institutionalised system of ideas about mathematics education, which are based on an out-moded philosophy that doesn't recognise cultural difference as significant in mathematics education. They have the possibility to mediate the system of ideas through their own involvement with groups and individuals in their network of 'colleagues' who share the tasks of acculturation. As Popkewitz implies, they have the opportunity to reconstruct the boundaries and by this means to extend the possibilities of acculturative education so that it is not just 'antagonistic acculturation' as Wolcott describes it.

The most powerful set of 'colleagues' in the school context are probably the parents of the students who, despite that fact, are usually excluded or at least marginalised from the formal educational encounter. In both chapter 5 and 6 we have seen some important evidence about ways in which teacher/parent 'power-sharing' can be a productive activity.

In chapter 5 the three case studies show us how powerless parents feel in their normal dealings with schools, but they also show us some of the possible actions that parents can take when they are aware of the transition situation their children are in. In Case Study 1, Kashif's parents are not sufficiently aware of the home/school differences in cultural and linguistic terms that seem to be the root cause of his difficulties at school. The parents are certainly enthusiastic about education, and about supporting their son's school education, but they lack the necessary information about the cultural nature of the differences to make a real positive contribution to his schooling.

In Case Study 2, Rachel's parents perceived and to some extent understood the cultural differences and transitions that she was experiencing. However this for them was a 'burden' and a worry:

M: I do get worried, and I've always said, that's why...I want to help her and I suppose that's why I probably don't do as much with her what I could....I wish there was courses, I know its hard for schools and that but I wish they would run courses for parents...and be shown exactly how...they're taught in school.

In Case Study 3, however, Jafar's parents were quite aware of the home/school differences but found successful strategies to use at home so as to minimise the negative effects of those differences. The concern as with Rachel's parents was to get more information from the school and to persuade the teacher to help them with their home mathematical activities.

Chapter 6 reports on a contrastive experience, namely where, by means of an imaginative intervention scheme, the teachers became more educated about the home mathematical activities and situation of their students. The teachers involved in this activity expressed many positive views about their involvement in the home contact work, although it clearly does not solve all the problems. For example:

I had known theoretically that yes, there is a lot of information out there. I am not working from a deficit model. But I still needed this first hand experience, even after I read all articles. I know their strengths now. But now that I know that from the home visits, how do I make sure that the child is aware of this, that they have all these experiences?

The following teacher expressed her views about the necessary knowledge that she felt she lacked in other ways:

I feel like sometimes I'm limited in my own knowledge as far as what I want to do mathematically. And so, I have to go to books and say, 'now, is that really where I want to go with my 4th and 5th graders? Or do I want to go in that direction? And would this be considered rigorous math? And will it work when my kids get tested on [a district standardised test]? Will they have learned something that will transfer over?' And that's threatening, really threatening.

However if that problem can be surmounted we can see from the following teacher's account that the intervention project achieved the necessary shift in thinking. Her testimony bears witness to the positive outcomes that both the home visits and the consequent shift of thinking can create:

'At first I thought it [going on a home visit] was very stinky. And then once I did the home visit, I was very impressed. ... [The student] turned around from a 'C' and 'B' student to an 'A' student. ... I was really shocked at the bond that was created. I never anticipated that nor did I anticipate the reception at the home. They were like 'come in, come in!' ... I think it is really powerful. I actually didn't want to do the interview. Because that was one of the requirements, I decided I would do it because I had to do it, but I didn't want to do it at all. I was very, very surprised at the results.'

The experience of the home visits was not just powerful for the teachers but also for the parents, as these comments show us:

'Engaging with our children in the mathematics, allows us to see them [our children] differently, that it is not sufficient to attend to all their other needs, but that it is important that we as parents have these types of discussions. We also realise that though we may not have a certificate in hand, we are also teachers.'

From all these interviews we can see how the 'intercultural' intervention and experience has made them both aware of a broader system of ideas and rules governing mathematical behaviour. In Popkewitz' terms the teachers particularly are now aware of the power of the system that up until then has normalised and constructed the rules governing their thinking about mathematics and mathematical behaviour. They are now in the process of reconstructing those rules and their perceptions of the knowledge within the conceptual structure of their classrooms that will develop power more productively, for the benefit of all their students.

What then of the learners' peers, the teacher's other potential 'colleagues' in the acculturation process? What role do and can they play in productive power-sharing during transition? Aware teachers seem to know how peer pressure influences learners, so in our terms an aware teacher/acculturator should know how peers can socially influence the reluctant 'inmate' or 'outsider'. And how might this kind of activity relate to the idea of the didactical contract that always exists between teacher and learner?

First of all let us see what can be learnt from the earlier chapters about the relationship between learners and their peers in the transition situation. To theorise this Gee's (1996) construct of 'borderland Discourses' will be used. These are: 'Discourses where people from diverse backgrounds, and, thus, with diverse primary and community-based Discourses can interact outside the confines of public-sphere and middle-class elite Discourses.' (p. 162) Borderland Discourses are those which take place in the 'borderlands' between the primary and secondary discourses. One's primary Discourse is learned within their family, and kinship groups, but a secondary Discourse is: 'a tradition passed down through time in ways people who 'belong' to the Discourse tend to behave now and have behaved in the past in certain settings' (Gee, 1992, p. 109).

In our context here the dominant secondary Discourse is argued to be what others refer to in the English-speaking context as 'the mathematical register of English' (Pimm, 1986). It is the 'official' formalised way to discuss mathematical ideas, and is the way that generations of mathematicians and mathematical students are assumed to have discussed mathematical ideas. It is the discourse which we know as 'mathematical discussions', and anyone who has successfully studied mathematics to a high level will be able to understand and contribute to these discussions.

The 'borderland' here is the mathematics classroom, as it lies between the family's world and the mathematician's world, and its common discourse is the borderland Discourse in focus here. Discussions and arguments about the didactical contract are part of this discourse, as is all the talk about procedures for checking answers, debates over group work seating, answers and scores on classroom tests, homework assignments, etc. These topics do not constitute the focus of mathematical argument, and would not be part of the secondary mathematical Discourse, but they are certainly part of mathematical **classroom** discourse, and as such they constitute part of the material of the borderland Discourse therein.

This discourse involves all the classroom participants, including the teacher of course, and must be learnt by any newcomer, just as one must learn the discourse of any group one joins, or wishes to join. Discourses do not however have fixed boundaries, and one would certainly expect any mathematics classroom discourse to contain elements of the secondary mathematical Discourse, particularly as the students grow in mathematical competence and sophistication. Teachers' speech will contain several elements of the secondary Discourse as they try to help the students develop their own discourse patterns. Equally the primary Discourse of family and home will also be present particularly in the students' speech and constructs. Nevertheless there is good reason to believe that the predominant discourse in the mathematics classroom is of the borderland kind.

For example, one of the words that frequently emerged in the interviews with the students in chapter 3 was 'work', with the prescribed mathematical activity in the classroom being perceived by them as having the nature of work. 'More' or 'less' work would seem to be an easily understood measure for these students doing their mathematical activities in the classroom. Dan articulated it this way, in comparison with what he was used to 'back home':

Dan: Well I learnt the basics in Georgia right, and not like here, they wouldn't let you talk in class, you would be sitting down by yourself and working, working, as we have a lot more work in Georgia than this, a lot more, I mean we get maybe one exercise here, there we get ten times as much.

Reg noticed the change in home-work demands when comparing a new with a familiar teacher:

Int: What about now, are you happy with your new teacher?
Reg: No. Don't tell her!
Int: So she is a bit different?
Reg: She is really different, like she doesn't explain as much but she gives you lots and lots of homework. Mr T explains it more and didn't give you as much.

'Work' is thus part of the lexicon of the mathematics borderland discourse with 'work-load' as the underlying construct. As Bishop said in chapter 3: 'The teaching approach, the curriculum, the strictness or otherwise of the teacher, the behaviour of the peers, all contribute and become blended into a picture of the 'mathematical work-load' i.e. how much you are expected to work in class.' Clearly this is determined to some extent by the teacher and her/his expectations, but it is also clear from the interviews that the peers exert a powerful influence here as well.

Tra, an immigrant student who wanted to get ahead with her work knew this:

Int: Who yells at you?
Tra: All the boys and girls as they don't want to work.
Int: And they say hey you are working too hard?
Tra: Yes.

She is not the only one who feels this pressure. It is particularly difficult for such students who are further ahead in their work because of what they did 'back home', who might think it is an advantage in the new context, to find conversely that they are ostracised by the peers:

Int: What's it like being in your class?
G: I don't know, they call me a square because I know more than normal kids. I don't like them calling me a square.

The word 'square' here illustrates rather well the distinction between the secondary mathematical discourse meaning and that of the borderland discourse!

O'Connor (1995), FitzSimons et al. (1996) and others, write about the workplace as a place of learning, as a kind of classroom. One of the images referred to in chapter 3 is worth recalling here: 'that of the 'union officials', who are the classmates who have negotiated the terms of the didactical contract for working in that classroom with that teacher who they see as the 'boss'...it is the 'old hands' amongst the classmates who are educating the newcomer about the rules of engagement, the level of work to be done, the extent of collaboration etc.'

However it is always open for the student to decide whether or not to 'join the union' and some decide not to. One can still choose one's friends. In Moscovici's terms one can choose who will be one's sources of influence, one's significant others:

Int: Some students I have talked to have felt pressure from other classmates when they do well, they get called a square.
Mar: I don't worry about that, my life is, sort of, like I don't live my life for those people, they are obviously not my friends if they say that. And if they are just joking I don't mind.

So the evidence from the borderlands under discussion in this book encourage one to offer as a construct, the idea of 'the mathematics classroom as a work-place'. In chapter 4 the *ardinas'* mathematics learning takes place literally in a work-place, and the following extract indicates the role of the experienced *ardina* Manu in the acculturation and transition process of the peers:

His main responsibility was in relation to the ardinas.... His first action was to write down a table with four columns: Taken, Sold, Left and Paid and each row was allocated to each ardina. This was his tool for control of the whole process and Manu made this record visible to all those who wanted to see it during the day. He updated the records in the case that some of the ardinas take more newspapers to sell, writing down again in a different row the name and the number of newspapers taken. This was also his record for the final account at the end of the day. Then Manu organised the money and the newspapers not sold in order to make the final account with Disidori...The ardinas recognised authority in Manu but they keep considering him essentially as an ardina and they had the perception that he had not substituted Disidori ('he is just helping him' as Kaka said) and that he was not a member of the staff of the newspaper agency. For them, Manu had a special status in respect to Disidori and so indirectly to the institution.

Manu is clearly one of the 'old hands' who is not only trusted by the 'boss' Disidori, but also by the 'newcomers'. His use of a spreadsheet functions both as a control but also as an explicit teaching and learning device. It makes the borderland mathematical discourse transparent and accessible to the learners in transition. O'Connor (1995) adds to this interpretation: '..in most workplaces workers quietly resist many official edicts and directives, as their own experiences tell them they won't work, or won't work as well as they can perform the task. Thus, workers employ 'unofficial' or 'theories in use' (or their specialised 'local knowledge') rather than the 'official' or espoused theories of the organisation' (p. 83). These unofficial theories in use form part of the borderland discourse and thus act also as learning tools for the newcomers.

For example in chapter 6, Civil and Andrade report comments from students A and B, who even at eight years of age have developed their own unofficial theories, and are able to use them to discuss the level of work that the teacher has determined for them:

A.: I hate math because I don't like low numbers, I like high numbers. They're harder and you get smarter. With low numbers, it's too easy. (8 year old)
B.: More really hard ones, because it's more fun to learn harder things. A lot of the stuff we already learned it in second grade.

In chapter 2 there is reported a significant incident with the student Miguel that shows us another part of unofficial theory, regarding the nature of mathematics. It is one that permeates much of the borderland discourse of classrooms with many immigrant students, because of the importance of the epistemology of mathematics in the educational institution.

Teacher: I want you to think, for tomorrow, a mathematical problem or situation that can be linked with this photograph (a picture of a rural market with a selling woman)
Miguel: (the next day) This was a trick! There is no mathematics problem, the woman has never been to school, she does not know mathematics.

Gorgorió, Planas and Vilella explain the response this way:

'Miguel is a 16 years old gypsy student, that works cleaning houses, and helps his family selling on the weekend street markets. He has a very low representation of his group regarding mathematical knowledge, he is sure that if his people knew mathematics they would not be selling in street markets. He does not accept that there is a need to know mathematics to sell in a market and, therefore, he can not see any mathematics at all in this practice of his community. He only accepts as mathematics the officially established curriculum within the school.'

Of course the irony about this example is that the teacher in this instance was trying to broaden the traditional construct of mathematics as described in the official curriculum to include mathematical practices in other contexts. She was doing this to facilitate their transition by developing the students' awareness of mathematical ideas in their everyday lives. Miguel however wants to keep the 'official' epistemological construct of mathematics in place.

This example also illustrates that these unofficial theories are an essential part of the negotiations regarding the didactical contract, which the learners and teachers simultaneously agree to and renegotiate throughout the teaching/learning encounter. In chapter 2 again Saima is unhappy about the mathematics she is learning and tries to renegotiate her didactical contract with the teacher:

Saima: Miss, I'm wrong in your class
Teacher: What do you mean? Why do you say that?
Saima: I do same mathematics as boys, but I do not same work ... I want not be a mechanic. Please, can I do the mathematics for girls?

So from the perspective of the learners in transition, the borderland discourse of the mathematics classroom has two simultaneous functions. On the one hand it is a discourse of the workplace, with its issues of contract, negotiating with the 'boss' and with the 'union' over work-loads, together with associated privileges, rewards and punishments, which have not been discussed here but which are exemplified in the earlier chapters. On the other hand it is a discourse for learning, with its unofficial theories, constructs and conceptual tools.

In relation to the power of the teacher, Gee's (1992) comments are interesting: 'The borderland discourse is a form of self-defence against colonisation, which like all organised resistance to power is not always successful but does not always fail' (p. 150). The teacher sees the borderland discourse predominantly in its function as a vehicle for learning, and in particular for learning the secondary, mathematical, discourse. The students' perception is more towards the second function, as a discourse of the mathematics classroom workplace, and particularly as a discourse for maintaining their primary, family, discourses.

How should the teacher who seeks to engage in cultural interaction then proceed? O'Connor's (1995) writing is again helpful here. He is concerned with workplace learning, and discusses Gee's ideas amongst others in connection with finding productive ways forward in that context. He argues: 'The point at which workers abandon the defensive aspects of the 'borderlands' will be when they see and are part of changes which they value and trust...This can be achieved by establishing approaches which share common values and objectives, collaborative 'partnerships and (which) establish ownership and control of ideas, planning and learning strategies as part of the dialogue and practices of the workplace' (p. 84).

The onus is clearly on the teacher not just to be aware of the role and importance of the borderland discourse for the students, but to pay continual attention to their workplace perspective of it, to recognise it as a genuine vehicle for the students' learning, and to see the value of allowing the involvement in it of both the students' primary discourses as well as the secondary mathematics discourse. Even if the mathematics classroom is the borderland, the boundaries of the borderland discourse are permeable and negotiable.

5. FROM INTENTIONAL ACCULTURATION TO CULTURAL PRODUCTION

In this chapter the focus has been on transition as an acculturation process and the people involved in that process. The concern has been to build on the data and analyses in the empirical chapters, and to elaborate three theoretical strands that draw on constructs from a variety of literatures that are seeking to explore new interpretations of the educational process.

The analyses in the chapter have raised some important issues, as well as indicating teaching implications, for the teachers and others involved in the acculturation process. Some of these have already been pointed out, but in this final section of the chapter the strands will be linked together. The framework for this section of the chapter comes from the idea of 'The cultural production of the educated person' which is the title of Levinson, Foley and Holland's (1996) book.

Their book illustrates the complementarity of 'structure' and 'agency' in that both have roles in the educational context, they are in some sense in opposition, and there is a constant interplay between them. As they put it: '..the very ambiguity of the phrase operates to index the dialectic of structure and agency. For while the educated person is culturally produced in definite sites, the educated person also culturally produces cultural forms' (p. 14).

This dialectic is at the heart of the acculturation process, and is vividly demonstrated by the various empirical observations in the other chapters. The educational structure is given by the formal/non-formal institution and its chief agents the formal/non-formal teachers, and from the acculturation perspective of this chapter the main issues concerned these teachers' power, and their use of it. Cultural reproduction theory envisages subjects as receivers of cultural messages without agency, with power as assumed to belong to the structure.

Agency in this context is with the learners, as in Willis' classic (1977) ethnographic study of working-class 'lads' in the north of England. In this chapter, and indeed in this book the learners have been shown to experience many cultural conflicts as they struggle with the power of the institutional structure, often in the form of the teacher and the teaching. Nevertheless they clearly have agency over their own acculturation.

Weiss in her Foreword to the Levinson, Foley and Holland (1996) book, puts the dialectic in another way that enables a further development to be seen. She says: 'The authors are clear that the processes of cultural production must be seen as lodged in that space between structure and agency…' (p.xi) In this chapter, as well as identifying the structure of institution and teacher, and the agency of the students, there has also been discussion of the 'borderland' of the mathematics classroom. Weiss' conceptual 'space between structure and agency' could be conceived of as the borderland Discourse space of the mathematics classroom.

This is the contested space where cultural production occurs. This is where the primary Discourse of the students' families and communities meets the secondary Discourse of the mathematical community through the teacher. It is where the co-construction of cultural meanings, values and practices develops.

This space is where Popkewitz' 'productive power' can be developed. People are not considered as 'possessors' of power (power as sovereignty in Popkewitz' words) but rather as developers of systems of knowledge and rules from which power can be derived. Teachers and students (together with other colleagues such as parents) can either actively sustain or recreate the existing systems of knowledge that allocate power in the already determined ways. Or they can together challenge explicitly the existing notions of mathematical knowledge and practice, and can co-construct alternative knowledge and practices. The mathematics classroom borderland Discourse is the both the vehicle and the source of the new knowledge.

This is also the space where explicit cultural interaction can be fostered, stimulating both cultural conflict and cultural consensus. Cultural conflict in the 'exclusive partitioning' sense of Valsiner and Cairns is not as productive a construct as the 'inclusive' sense of being the complementary partner of consensus. Mathematics teaching and learning that recognises both conflict and consensus as significant in the learning process, is likely to be a more significant pedagogy in the context of cultural difference. The borderland Discourse offers the opportunity for the sharing and comparing of ideas and constructs.

This chapter ends with two quotations. The first illustrates the issue of learning to trust the teacher, which has been fundamental yet tacit throughout this discussion. Trust is associated with openness on the part of the teacher, which is a necessary part of making explicit the cultural interactions and the productive power-sharing. Tra in chapter 3 hints at this trust and indicates how important it is from the perspective of the immigrant student:

Int: How good do you think your teacher thinks you are?
Tra: I don't know, no idea. It is hard to know what the teacher thinks.
Int: Tell me why it is hard to know.
Tra: Even though you might do well they might think something different about you.

Finally Wolcott's (1974) experience to which we have already referred is instructive also. This quotation cautions against the supposed optimism of the idea with which we began this chapter 'intentional acculturation'. He was 'white' but was teaching an entire class of Kwakiutl students. In discussing more appropriate teaching approaches he argues: 'The teacher may feel a greater need to alert his pupils to the fact that he has *not* been able to provide them with all the prerequisite skills for successfully 'passing' in the teacher's own society than to fill them with hopes and promises which few may ever realise' (p. 147).

Cultural production should be the goal, not intentional acculturation.

REFERENCES

Abreu, G. de (1995). Understanding how children experience the relationship between home and school mathematics. *Mind, Culture and Activity: An International Journal,* 2(2), 119–142

Abreu, G. de, Bishop, A.J. and Pompeu, G. (1997). 'What children and teachers count as mathematics'. In T. Nunes, P. Bryant (Eds.), *Learning and teaching mathematics. an international perspective,* (pp. 233–264) Psychology Press, Hove.

Barth, F. (1969). *Ethnic groups and boundaries.* Boston: Little, Brown and Company.

Barton, B. (1996). Anthropological perspectives on mathematics and mathematics education. In A.J. Bishop, K. Clements, C. Keitel, J. Kilpatrick & C. Laborde (Eds.), *International handbook of mathematics education.* (pp. 1035–1053) Dordrecht: Kluwer.

Becker, E. (1962). *The birth and death of meaning.* New York: The Free Press

Bishop, A.J. (1988). *Mathematical enculturation: a cultural perspective on mathematics education.* Dordrecht: Kluwer

Bishop, A.J. (1994). Cultural conflicts in mathematics education: developing a research agenda. *For the learning of mathematics.* 14, 2, 15–18

Coombs, P.H. (1985). *The world crisis in education: the view from the eighties.* New York: Oxford University Press.

FitzSimons, G., Jungwirth, H., Maass, J. & Schloeglmann, W. (1996). Adults and mathematics (adult numeracy) In A.J. Bishop, K. Clements, C. Keitel, J. Kilpatrick & C. Laborde (Eds.) *International handbook of mathematics education* (pp. 755–784). Dordrecht: Kluwer.

Gee, J.P. (1992). *The social mind: language, ideology and social practice.* New York: Bergin and Garvey

Gee, J.P. (1996). *Social linguistics and literacies: ideologies in discourse.* London: Taylor and Francis

Kirshner, D. & Whitson, J.A. (1997). *Situated cognition: social, semiotic, and psychological perspectives.* New Jersey: Lawrence Erlbaum Associates

Levinson, B.A., Foley, D.E. and Holland, D.C. (1996*). The cultural production of the educated person.* Albany, NY: State University of New York Press

McDermott, R.P. (1974). Achieving school failure: an anthropological approach to illiteracy and social stratification. In G.D. Spindler (Ed) *Education and cultural process: towards an anthropology of education.* (pp. 82–118) New York: Holt, Rinehart and Winston.

Moscovici, S. (1976). *Social influence and social change.* London: Academic Press

O'Connor, P. (1995). Workers' cultures and learning: a spanner in the works. *Critical Forum*, 4, 2, 70–102

Pimm, D. (1986). *Speaking mathematically: communication in mathematics classrooms.* London: Routledge and Kegan Paul.

Popkewitz, T.S. (1999). Introduction: critical traditions, modernisms, and the 'posts'. In T.S. Popkewitz, & L. Fendler (Eds) *Theories in education: changing terrains of knowledge and politics.* (pp. 1–13) New York: Routledge

Restivo, S. (1993). The social life of mathematics. In S. Restivo, J.P. van Bendegem & R. Fischer (Eds) *Math worlds: philosophical and social studies of mathematics and mathematics education.*(pp. 247–278) Albany, USA: State University of New York Press.

Shantz, C.U. & Hartup, W.W. (Eds) (1992). *Conflict in child and adolescent development* Cambridge: Cambridge University Press.

Spindler, G.D. (Ed) (1974). *Education and cultural process: towards an anthropology of education.* New York: Holt, Rinehart and Winston

ValsinerJ. & Cairns, R.B. (1992). Theoretical perspectives on conflict and development. In C.U. Shantz & W.W. Hartup (Eds) *Conflict in child and adolescent development* (pp. 15–35) Cambridge: Cambridge University Press.

Willis, P. (1977). *Learning to labor: how working class kids get working class jobs.* New York: Columbia Press.

Wolcott, H.F. (1974). The teacher as an enemy. In G.D. Spindler (Ed) *Education and cultural process: towards an anthropology of education.* (pp. 136–150) New York: Holt, Rinehart and Winston.

CHAPTER 9

SHIFTS IN MEANING DURING TRANSITIONS

NORMA PRESMEG
Illinois State University

'The presumptions of meaning are based on community, purpose, and situation. It is futile to discuss the meaning of a word or term in isolation from the discourse community of which the speaker claims membership, from the purpose of the speaker, or from the specific situation in which the word was spoken. Indeed, it is not the word that has meaning, but the utterance' (Clarke, 1988, pp. 99–100).

1. THEORETICAL LENSES

At the heart of the experiences of transitions between contexts which are the focus of this book, lie subtle and not-so-subtle shifts in the meanings that participants construct for these experiences. Meanings entail the making of connections. This chapter analyses in particular some of these shifts in the connections transitioners construct, especially as these connections pertain to the learning of mathematics. The theoretical framework of this analysis is of necessity fairly wide. As the opening quotation suggests, the meanings given to mathematical constructs and the symbolisations that may accompany them are intricately bound together with sociocultural meanings, values, and beliefs of students in transition experiences. Some of these elements will have been formed in prior situations that may bear little resemblance to the current situation of a student – but that are nevertheless present in the shifting tides of thought as that student endeavours to come to terms with, and to give new meaning to, the current situation and its mathematical elements. Even if a physical move (e.g., from one country to another or from home to school) is part of the experience, the major components in all transitions are mental ones by virtue of the need to construct new meanings. It is these mental shifts in connections that this chapter examines both from an individual point of view, resonating with a cultural psychological perspective (Abreu, chapter 7), and from an anthropological one embracing acculturation and resolutions of repeated cultural conflicts (Bishop, chapter 8).

Clarke's opening quotation is relevant to a second aspect of the theoretical framework of the present chapter. That it is not 'the word' that has meaning, but the utterance, was powerfully demonstrated by Bourdieu (1995). Clarke derived his constructs from theoretical notions of Bakhtin (Clarke, 1988, p. 100), but it is

G. de Abreu, A.J. Bishop, and N.C. Presmeg (eds.), Transitions Between Contexts of Mathematical Practices, 213–228.

formulations of Bourdieu, including his constructs of *cultural capital* and *symbolic violence*, which are the principal lenses used to explore some of the issues involved in shifts of meanings during transitions in this chapter. Bourdieu's analytic tools include the notions of *habitus* and *symbolic power*, and show the inadequacy of a Saussurean structural analysis to cope with the shifting meanings in human communication (Bourdieu, 1995, Grenfell & James, 1998).

Because his constructs are the main theoretical lens used in this chapter, the relevance of Bourdieu's ideas to the present purpose will be elaborated in a short digression here. This relevance consists in 'the innumerable and subtle strategies by which words can be used as instruments of coercion and constraint, as tools of intimidation and abuse, as signs of politeness, condescension, and contempt' (Bourdieu, 1995, p. 1, editor's introduction). The experiences of learners in transition are often subject to these manifestations of power and social inequality. Bourdieu uses the ancient Aristotelian term *habitus* to refer to *a set of dispositions which incline agents to act and react in certain ways* (Bourdieu, 1995, p. 12). An individual's durable and enculturated habitus gives a sense of how that individual should act and respond in daily life. According to Bourdieu there is not only a 'practical sense' associated with habitus, but also a physical or bodily set of habits, a 'bodily nexus' inculcated from childhood, structured to reflect the social conditions in which the dispositions were acquired. 'Sit up straight', 'Don't talk with your mouth full', etc., are social injunctions that engender a certain habitus. The durability of habitus is seen in perceptions that structured dispositions of this kind 'are ingrained in the body in such a way that they endure through the life history of the individual, operating in a way that is preconscious and hence not readily amenable to conscious reflection and modification' (Bourdieu, 1995, p. 13). Finally, habitus is generative in the sense that even in fields other than the original source, it tends to produce 'practices and perceptions, works and appreciations, which concur with the conditions of existence of which the habitus itself is the product' (p. 13).

The foregoing formulation of habitus points to the magnitude of the changes required of immigrant children and their families in transition experiences. What is being asked of them is the construction of a new set of acculturated dispositions, in situations where they may not even understand *why* the dispositions of their old habitus are failing them. The enormity of the task becomes clear.

These issues are elaborated in Bourdieu's notions of *cultural capital* and *linguistic capital*. He describes various forms of capital, for instance, *economic capital* (material wealth), *cultural capital* (knowledge, skills, and other cultural acquisitions, as exemplified by educational or technical qualifications), and *symbolic capital* (accumulated prestige or honour) (Bourdieu, 1995, p. 14). *Linguistic capital* is not only the capacity to produce expressions that are appropriate in a certain social context, but it is also the expression of the 'correct' accent, grammar, and vocabulary. The *symbolic power* associated with possession of cultural, symbolic, and linguistic capital, has a counterpart in the *symbolic violence* experienced by individuals whose habitus is devalued.

Symbolic violence is a sociological construct. In that sense it is a powerful lens with which to examine actions of a group and ways in which certain types of knowledge are included or excluded in what the group counts as knowledge. Examples of this 'valorisation' or lack thereof abound in Abreu's (1993) research in rural Brazil. However, this notion of symbolic violence leaves a theoretical gap relating to the ways in which individuals then choose to construct, or choose *not* to construct, particular meanings – mathematical or otherwise. The notion of *Discourses* formulated by Gee (1992), and in particular his extension of the construct to *borderland Discourses*, those 'community-based secondary Discourses' situated in the borderland between home and school (Gee, 1992, p. 146), resonate with Bourdieu's ideas of habitus and symbolic violence. These ideas go some way towards closing this theoretical gap by raising some issues of individuals' choices (see also Bishop's use of Gee's constructs in chapter 8 of this volume). Gee's work was in the context of second language learning but it is also useful in the analysis of meanings given to various experiences by mathematics learners in transition situations. These ideas are elaborated in subsequent sections. I believe strongly, with Cobb and Yackel (1995), that it is important to consider both sociocultural issues and individual psychological ones in attempting to understand the complex processes involved, either directly or indirectly, in the learning of school mathematics.

These lenses, then, are eminently suitable for looking at some of the shifts in meanings constructed by mathematics learners in transition experiences. The processes involved may be turbulent ones. This chapter examines evidence from the empirical chapters in this book that mathematical meanings are intertwined with social meanings as learners attempt to change the dispositions that constitute habitus in these processes.

A few further theoretical notions which are relevant to these processes are the constructs of intersubjectivity (Lerman, 1996), and positioning (Walkerdine, 1982, 1988, 1990, 1997). These constructs relate to communication and to power structures respectively, and both may cast light on why individuals are motivated to change their meanings during transitions. Intersubjectivity relates to the interpersonal and social aspects of human communication: to be accepted in a group, an individual must have sufficient cultural and linguistic capital for intersubjectivity to occur. This entails the sense that meanings are shared, at least to some extent. The sense of shared meanings is also a component of the symbolic power afforded by the acquisition of cultural and symbolic capital, as illustrated in chapter 3 when some of the immigrant students described there started to 'shout back' when others in the group called them names. Without this intersubjectivity, the opportunity for continuing symbolic violence is strong. In the following analysis, I shall not use the poststructuralist notion of positioning extensively; however, this construct is useful in explaining how a person's situation in the hierarchy of a group's power structures (that person's positioning) may provide the impetus for changed meanings. As an individual's positioning changes, both the individual and other group members are called upon to re-evaluate the meanings implicit in the symbolic power (or lack of it) attributed to the individual by the group. This in turn may change the positioning of the individual, and so this reflexive process may be an integral part of the change of meanings during transitions.

One further viewpoint that substantiates the decision to take both an individual and a sociocultural perspective in the analysis of transitions in this book, is that of

Pimm (1995), who wrote in terms of 'mathematics deriving from both inner and outer experiences, and meaning as being generated in the overlapping, transitional space between these two powerful and sometimes competing arenas' (p. 11). Pimm's viewpoint resonates with the borderland Discourses of Gee (1992) which may impact this transitional space. The intertwining of learners' inner and outer experiences in the construction of meanings during transitions is a theme of this chapter.

Although there is overlap, the following sections are organised around meanings connected with the difficult and varied transitions of immigrant children in mainstream schools, the role of parents in construction of mathematical meanings in these situations, and the roles of language and associated symbolism during the implicated transitions. In all of these sections the evidence from data provided in this book suggests that mathematical meanings are subordinate to other issues, such as the meaning of a student's personal identity in the group, while that student is in the midst of a transitioning process.

2. IMMIGRANT LEARNERS IN MAINSTREAM SCHOOLS

I want to begin with a brief note about the interpretation of meanings in the empirical chapters of this book. Writing about the meanings of other people (and, indeed, of one's own) of necessity involves interpretation. There is a choice about what elements should be foregrounded, and what it is necessary to report as background to the interpretation (Gee, 1992). My initial interpretations of meanings of participants in the transitions involved in this book were subjected to the scrutiny of the respective authors, whose own interpretations in their chapters were then taken into account to cast light on the processes involved in various shifts of meaning.

In chapter 2, Gorgorió, Planas, and Vilella wrote that immigrant children 'need to build a bridge from meanings of their initial situation to the present one.' They further suggested that the school has a role in this process in contributing to the creation of continuity between the home and the host country's cultural meanings. Whether it is possible for children to experience continuity of meanings in this bridge-building process is an open question that raises complicated issues, and certainly if it is possible, the process will be subject to many influences, of which the school is a significant one.

In many ways, Saima's remarks (chapter 2) suggest a paradigm case embodying the cultural conflicts that confront many immigrant children. In expressing her belief that the mathematics she was doing was more suited to boys and therefore not right for her ('I do not want to be a mechanic'), Saima is describing a discontinuity between the cultural values of her home, with concomitant beliefs about the roles of girls, and the meanings she perceived to be connected with school mathematics. Conflicting roles in her home and in the school – in fact, her identity and self-image – may be involved. The fact that she tells her teacher about this cultural conflict suggests that she is attempting to bridge the gap in meanings for her roles at home and in the school mathematics classroom. There are several possibilities. One might consider Saima's options in terms of the three standpoints used by Gorgorió et al. in chapter 2, namely, as an individual, as a member of the classroom community, and as an individual with a sociocultural identity. These three viewpoints taken by the authors of chapter 2 evolved from their consideration of the 'emergent perspective' of Cobb and Yackel (1995, p. 6), in which a social and a psychological perspective were balanced. The former perspective, consisting of the negotiation of classroom social norms, sociomathematical norms, and classroom mathematical practices, has its counterpart in the latter, consisting respectively of individual students' 'beliefs about their own role, others' roles, and the general nature of mathematical activity at school', mathematical beliefs and values, and mathematical conceptions. Here the marrying of individual and social perspectives will be taken, in line with Gorgorió et al.'s analysis, to consist of an individual student's perspective, a perspective as a member of the classroom community, and a perspective as an individual with a sociocultural identity. There is of course overlap between the categories: the self-identity of the student belongs with the perspective of the individual, the perspective as a class member, and is at the same time intricately bound up with a sociocultural identity which embraces both of the former categories. Nevertheless there are differences in the elements that are foregrounded in the three views.

Firstly, *as an individual*, on the one hand, Saima might resolve her home-school conflict by rejecting the school notion of mathematics as gender-free and suitable for all. This option could lead, in the extreme case, to Saima's withdrawal from participation in the sociomathematical norms of the classroom (Cobb & Yackel, 1995), and even to her rejection of the institution of school itself (Valdés, 1998). On the other hand, she might reject the values and beliefs about roles of girls in her home culture. Extreme consequences of this option are reminiscent of the 'Education for extinction' of the American Indian boarding school experience (Adams, 1998). Either of these extremes would result in personal meanings that would rob her of the richness a bridging of the two cultures might provide (Nieto, 1996). Nieto's analysis of cases of students with positive bridging experiences revealed that they had all found ways to construct meaningful identities that incorporated participation in both home and school cultural practices. Withstanding of peer pressure and avoidance of situations that evoked such pressure were crucial elements in this process: students needed the courage to *be* different.

One noteworthy point is that if the sociocultural norms are rigid, Saima's choice of values is an either-or option. The law of the excluded middle holds here: one either pets the cat or one eats it (Presmeg, 1988). However, in the bridging experience it is possible, without conflict, to keep different values in different mental compartments, to be used in the appropriate different sociocultural contexts. There is evidence (e.g., in Spindler, 1974) that individuals can and do handle this compartmentalization of values without harm to their personality structures. 'At home I add, at school I multiply,' said Bishop's Papua New Guinean interviewee simply, when confronted with conflicting choices of ways of finding area in these two sociocultural contexts. However, as Bishop (chapter 8) describes, based on the work of Valsiner and Cairns, there is another way of viewing cultural conflict. The either-or choices are an example of *exclusive partitioning*. On the other hand, Saima may confront the dichotomy of 'girls' mathematics versus boys' mathematics' and through her agency try to change the perceived norms either of her home or of the classroom, perhaps succeeding in eliminating the dichotomy. Thus her choice may not be the extant either-or option of meanings (although her words do seem to indicate that she has accepted the exclusive view). Negotiation appears to be a factor in constructing the *inclusive partitioning* that 'preserves heterogeneity and enables conflict to be defined in terms of the nature of linkages between the differentiated parts of a whole' (Bishop, chapter 8). Saima's comments indicate negotiation with her teacher; she might also negotiate with her parents if she chooses to engage in what she currently views as boys' mathematics. But the main element that would enable this sociocultural negotiation would be a transitioning of her own mental meanings past this dichotomy.

Following on from these considerations, in the second of Gorgorió et al's categories, viewing the immigrant student *as a member of the classroom community*, the student's choices may be mediated by the meanings attributed to the situation by the teacher. In Bourdieu's (1995) terms, the teacher and her mainstream students have the cultural capital, in classrooms that contain immigrant students. The potential for symbolic violence is strong in this situation – as illustrated by Gorgorió et al. in the striking image of the placement of the desks of immigrant children in positions where the teacher will not have to deal with them very often. The meanings that these immigrant students construct for this marginalisation, subtle as it is, cannot but help influence their choices regarding level of participation in negotiating the social norms and the sociomathematical norms (Cobb & Yackel, 1995) of this classroom. Thus the cycle of marginalisation may continue. A 'logical' choice, from the point of view of an immigrant student in this situation, might be to drop out of school as soon as possible, as Lilian did in a similar situation described by Valdés (1988) – only to find that this choice did not solve wider problems of marginalisation and symbolic violence in the society. In many immigrant homes, dropping out of school is not an option: parents may view schooling as the road to increased cultural capital for their children (Civil & Andrade, this volume; Grenfell & James, 1998, p. 66).

Continuing the theme of the 'invisibility' of immigrant students in some classrooms, what are the meanings that immigrant students construct when not only their cultural capital, but in some cases, their very presences, are ignored by a teacher? Nieto (1996) describes the case study of a Lebanese student, James, in Springfield, Massachusetts, in the USA. Lebanese immigrants constituted an 'invisible minority' in this community. After he had described his exclusions from the foreign language month of his school, the multicultural cookbook, and the multicultural fair, his ambivalence is apparent in the following comment:

> 'I don't mind, 'cause I mean, I don't know, just, I don't mind it .... It's not really important. It is important for me! It would be important for me to see a Lebanese flag. But you know, it's nothing I would like enforce or like, say something about.... If anybody ever asked me about it I'd probably say, 'Yeah, there should be one.' You know, if any of the teachers ever asked me, but I don't know....' (Nieto, 1996, p. 165).

In chapter 3, Bishop described the case of Tra, who commented, 'like sometimes even though you are good, you are kind of like a shadow'. This striking analogy was her way of expressing the meanings associated with her invisibility the previous year when she was 'so quiet as it was depressing as no-one noticed you'. However, as Gee (1992) noted, 'episodes are not in the physical happenings of the world, which are continuous, nor are they in the head, which contains as its store of information a network of diverse associations – *they are in the telling*' (p. 60, his emphasis). The point is that both Tra and James have a choice to make in how they construct their invisibility, and how they act on it. What they choose to do will impact the 'ever-growing and changing history of the Discourse' (Gee, 1992, p. 109). In order to try to change the Discourse, students need to be aware that this is a possibility, and even when they are aware they may choose not to challenge the status quo, as in James's case. By way of contrast, when Tra's peers now call her 'nerd head' because they consider she is working too hard, she currently 'just yell(s) back and start(s) to annoy them'. By resisting the social norms expressed by her peers, she may in a small way change the 'borderland Discourse' of the classroom (Bishop, chapter 8). The complementarity of structure and agency in negotiating the classroom social norms is illustrated in Tra's case. However, this complementarity is a complex and delicate balance, and how immigrant students will construe it depends in part on the meanings they construct for these classroom episodes.

Thirdly, the immigrant student is an *individual with a sociocultural identity*. As Gorgorió et al (chapter 2) point out, students attach meanings to situations, to actions, to themselves, and to other people through an interpretive process. This process is constantly revised and adumbrated through the acquisition of new experiences. Intersubjectivity is a crucial element in the transitioning of immigrant students, through this interpretive process. Bridges are constructed successfully to the extent that meanings are *perceived* to be shared, whether or not the deeper significances of these meanings are compatible. Teacher affirmation of diversity (Nieto, 1996) may encourage the communications that nurture such perceived congruencies of meanings. As Bishop (chapter 8) points out, conflict *and* consensus are necessary elements in the inclusive view of a heterogeneous classroom. As is suggested by the research evidence in most of the chapters in this book, in the complexities of these transitioning processes mathematical meanings may be totally subordinate to sociocultural ones. The gaining of mathematical capital may be less pressing than the gaining of linguistic and cultural capital for an immigrant student in a mathematics classroom.

Many of these issues are continuing themes in this chapter.

3. THE ROLE OF PARENTS IN CONSTRUCTION OF MATHEMATICAL MEANINGS

In their study of Pakistani and English mathematics learners in England, Abreu, Cline, and Shamsi (chapter 5) show clearly the impact of parents' value judgements on the mathematical learning of their children. But the account also highlights the way in which value judgements of children in this regard impact the process. The vivid case study of Kashif in the second year of an English school is an illustration. The meanings constructed by his Pakistani mother do not resonate with his own constructed meanings with regard to the power structures and his own positioning in his mathematics classroom, as a 'Pakistani immigrant' who speaks Urdu with his parents at home. The language difficulties Kashif experiences in school, for instance in his reading, while identified as a factor by the researchers, seem to have been overlooked by his teacher, his parents, and by Kashif himself. With the glossing over of the issue of linguistic capital, Kashif appears to have a strong idea of the power afforded by cultural and symbolic capital in the person of his teacher. It appears that mathematical meanings are subordinate to sociocultural ones in this case too.

On the one hand, it is ironic that the meanings Kashif constructs for the symbolic power of the teacher cause him to devalue the help provided by his mother, who does not perceive the lack of resonance in their meanings, and hence does not understand the urgency of his school situation. 'But he still doesn't seem to make the progress', reports his teacher (chapter 5). In contrast, both his parents are of the opinion that his achievement in mathematics 'is improving'. On the other hand, an instance of where meanings of Kashif and his parents *do* resonate, is in the value attached to 'being good at mathematics' for the sake of getting a decent job in the society. This valorisation reinforces the symbolic power afforded to the teacher, hence devaluing the cultural capital of Kashif's parents and perpetuating a spiral of symbolic violence in which Kashif is caught. In order to break from this spiral and complete the transition process (which does not even exist as far as Kashif's parents are concerned, as Abreu et al. point out), Kashif's meanings will probably need to shift towards a valorisation of the cultural capital he and his parents *do* possess (Nieto, 1996). Greater communication between the teacher and his parents could probably facilitate a move in this direction. It is both a positive and a negative feature that Kashif's mother does not exhibit the lack of self confidence shown by mothers in similar situations in Reay's (1998) study. Greater awareness of the need for transition might bring a corresponding drop in confidence with the perception that her cultural capital has been devalued.

From an English home, by way of contrast, Rachel's case illustrates how resonance between perceptions of the value of the cultural capital of her family and that of the school classroom may create another set of difficulties. The transition involved here is from mathematical methods that were current when Rachel's mother attended school, to those that are taught by Rachel's teacher in the present day. This is an 'encompassing transition', in Beach's (1999) terms, because it takes place within the boundaries of one practice. Rachel's mother has an acute awareness of the need for this transition, if she is to help her daughter with her learning of mathematics. In this sense it is the mother's transition as much as the daughter's. Difficult and burdensome as the transition may appear to be, their awareness of the need for the process avoids the spiral of symbolic violence in which Kashif is caught.

On the surface, the third case study in chapter 6, that of Jafar, appears to be similar to that of Kashif, at least in the home backgrounds of the students, and in the use of Urdu as a home language, although both these students preferred to speak English with their siblings. Then what is it that helps Jafar to avoid the spiral described in Kashif's case, and to be considered a successful mathematics student by his teacher? In the first place, Jafar's mother is aware of the cultural transition involved. She has worked in the English school system and has gone to great lengths to find out what mathematical content and methods are involved in Jafar's school experience. One might say she has acquired sufficient cultural, linguistic, and symbolic capital to play a role in Jafar's transition. This set of dispositions, or *habitus* that his mother is developing, helps Jafar to construct positive meanings for his own transition process. He does not devalue his own home cultural capital, but sees it as underpinning his efforts to attain the symbolic capital of success and appreciation by his mathematics teacher. I am reminded again that all the successful transition experiences described by Nieto (1996) involved a similar affirmation of the home culture.

Civil and Andrade, in chapter 6, continue the theme of transitions between home and school, but with a more overtly pedagogical goal. In exploring ways to connect the home experiences of students of Mexican descent in Arizona with the mathematics they learn in school, these authors ask, 'Can we bring to the classroom the child's perception of him/herself as a learner outside school?' Their chapter elaborates a theme of this book, namely, that the identities of learners are involved in transitions across contexts in the learning of mathematics.

When ten-year-old Alberto does not adapt to the Arizona school system as well as his six-year-old and thirteen-year-old brothers do when the family emigrates from Mexico, a visit to their home by the authors of chapter 6 provides some illuminating insights. In Mexico, Alberto had successfully managed part of the family's bakery business. Unlike his brothers, he did not want to emigrate to the USA, and he resented his family's decision. Now his brothers are learning English, entering their transition processes, acquiring linguistic and cultural capital in the new environment. Alberto does not want to be any part of this. The meanings he constructs for the whole experience of coming to the USA are, initially at least, negative ones. One wonders whether a teacher's attempt to incorporate a simulated 'bakery experience', which would valorise Alberto's cultural and symbolic capital, in the classroom learning of mathematics, might help to facilitate Alberto's acceptance that a transition process is necessary if he is to remain and become a useful member of his new society.

In her *Funds of Knowledge* project, Civil has been facing and exploring solutions to issues such as this for several years (Civil, 1995). In rejecting a deficit model of learning for minority students in the USA, she and I share some concerns about values and beliefs held by teachers and students with regard to the nature of mathematical knowledge both in and out of school. In order to bring experiences that are authentic to students' lives into the mathematics classroom, it is necessary to challenge both students' and teachers' beliefs and values concerning the nature of mathematics (Presmeg, 1996, 1998).

The parents of immigrant students described in this book had generally high expectations for the performance of their children in mathematics. Parents may be sources of both pressure and support (Bishop, chapter 3). However, there is a poignant underlying issue which may influence the meanings that children attach to their parents' encouragement to do well in mathematics so that they may get a good job. Gor, (Bishop, chapter 3) has a father who taught him decimals, algebra and problem solving. Gor says that mathematics is one of his favorite subjects, and he is 'good at it'. When the interviewer asks if Gor's father is a mathematics teacher, it emerges that Gor's father works as a school cleaner in Australia. In Macedonia, 'He was a building technician for bridges, he had to test them', as Gor reports. There is a subtle contradiction in the messages here. The mathematical capital of Gor's father was of no avail in helping him to attain a job consonant with his qualifications, without the linguistic and cultural capital valued in Australia. The symbolic capital of his mathematical knowledge was in fact subjected to symbolic violence in his new country. His father's situation does not seem to have impacted Gor's mathematical experience negatively: in fact his meanings appear to be totally positive. However, this situation suggests just how subtle and complex the messages conveyed to immigrant students may be. Language and communication are a theme in the next section.

4. THE ROLES OF LINGUISTIC AND SYMBOLIC CAPITAL DURING TRANSITIONS

Researchers such as Ellerton and Clarkson (1996) have confronted the complexities of moving between two languages, as these transitions are experienced in mathematics classrooms. Language factors in mathematics teaching and learning have been explored quite comprehensively by these authors, who came to the conclusion that social, cognitive, cultural, linguistic, and affective factors affect communication patterns in mathematics classrooms. In this account, Ellerton describes a model in which the construction of meaning in mathematics is nested at the center of widening language categories including 'the specific domains of the language of mathematics', 'the language of the classroom', and 'real-world language' (p. 990). It is clear that an analysis of shifts in meanings during transitions will need to take these widening language categories into account.

In the intricacies of teachers' encouragement of student mathematical discourse, as Moschkovich (1999) points out, 'Two important functions of productive classroom discussions are uncovering the mathematical content in student contributions and bringing different ways of talking and points of view into contact' (p. 11). Meanings that students attribute to their classroom discourse may or may not resonate with the mathematical content that their teacher sees in it. The nested quality of the meanings implies that the languages of mathematics, the classroom, and the real world may all impinge on and contribute to these meanings. In the case of the non-English speaking background (NESB – for clarification of this classification and associated problems, see chapter 3) students that Ellerton and Clarkson describe, the range of possible connections is at once widened and curtailed, making the construction of mathematical meanings that much more difficult. The possibilities are widened by associations that belong to another language and culture; they are curtailed by the potential for symbolic violence. Lacking the linguistic and cultural capital of students for whom English is the first language, NESB students may overlook the potential provided by their wider associations. This aspect may explain to some extent the threshold effect, based on Cummins' work, which Ellerton and Clarkson describe. It may require a certain level of competence in a student's first language for that student to see the widened associations as a form of cultural capital that may be exploited in the learning situation. This view resonates with the notion of cultural conflict expressed as complementarity underlying explicit cultural interaction (chapter 8).

In several instances in the empirical chapters of this book, there was evidence of a lack of communication, or a breakdown in communication, between a mathematics teacher and an immigrant student. Several immigrant students in chapter 3 expressed their doubt about how their mathematics teacher might be constructing their ability. Tra said, 'Even if you do well they may think something different about you.' It would appear that interpreting body language and similar subtle signs is also a form of linguistic capital necessary in successful transitions in a mathematics classroom. Together with this is the need for the cultural capital to construct appropriate meanings for the linguistic and body communications. As Bishop (chapter 3) points out, immigrant students in transition experiences may be more aware than most of 'the need to understand the subtle signals that trigger teacher perceptions and determine teacher behaviours towards them, and ultimately teacher assessments of them.'

In an attempt to understand more about the meanings transitioners construct as a result of communication experiences, I want to consider more closely Gee's (1992) theoretical lens of Discourses. He explains, 'I use the term Discourse (with a capital D) to make clear that I am talking about something that means (discourses) though it is not in any one person's head, but rather an amalgam of language, bodies, heads, and various props in the world' (p. 87). He elaborates that Discourse involves 'ways of acting, interacting, talking, valuing, and thinking, with associated objects, settings, and events, which are characteristic of people whose social practice it is' (p. 91). Between the primary Discourse of the home and the secondary Discourse of the practices of the mathematics classroom, lies the borderland Discourse of interactions with peers (see also chapter 8). In the following passage, Gee (1992) continues to expand these lenses in a way that suggests to me that Discourses underlie habitus and Bourdieu's various forms of capital – not only linguistic, but cultural and symbolic capital too.

> 'Each Discourse apprentices its members and 'disciplines' them so that their mental networks of associations and their folk theories converge towards a 'norm' reflected in the social practices of the Discourse. These 'ideal' norms, which are rarely directly statable, but only discoverable by close ethnographic study, are what constitute meaning, memory, believing, knowing, and so forth, from the perspective of each Discourse. The mental networks in our heads, as well as our general cognitive processing abilities, are tools that we each use to get in and stay in the social 'games' our Discourses constitute' (Gee, 1992, p. 141).

Gee states elsewhere (p. 108) that these norms are in fact where the *meanings* of the acts, beliefs, values, and words of the group reside.

The foregoing formulation of the role of Discourses bridges the sociocultural and the psychological aspects of meaning, the perceived outer and inner worlds of individuals, and also makes clear the subtleties of interpretation necessary in the acquisition of new cultural, linguistic, and symbolic capital during transitions. Although each case is individual, Tra (chapter 3) in the experience of wondering whether her teacher values her work, Kashif (chapter 5) devaluing his mother's cultural capital in favour of his teacher's, Alberto (chapter 6) longing for the sense of worth he experienced in his role in the family bakery in Mexico, have all not yet constructed meanings that will enable them fully to enter the secondary Discourses of their mathematics classrooms. It seems likely that some borderland Discourses with peers will facilitate this meaning-making process during transitions, as in the case of Tra, who appears to have stopped 'being a shadow' – or feeling invisible – during her second year when she 'yell(s) back' at her peers. It is only when Tra takes upon herself the *agency* to be whom she wants to be in the classroom, that she is in a position to influence the borderland Discourse and start entering the secondary Discourse, which will signal a successful transition.

Outside the classroom, another sort of borderland Discourse, just as influential in a transitions process, takes place amongst the *ardinas* who sell newspapers in Cabo Verde. In chapter 4, Santos and Matos adopt Lave's view of learning as participation in communities of practice, rather than property acquisition. Bourdieu's notion of cultural capital appears to partake of both of these metaphors: acquisiton of cultural capital pertains to the latter, property acquisition, but enables the former, participation in communities of practice. The ethnographic study described in this chapter has the purpose of searching for an understanding both of how *ardinas* deal with the mathematical aspects of their practice, and of how they connect school mathematics with this practice. There is an implicit transitioning back and forth between school mathematical practices and the mathematics inherent in the practice of selling newspapers. The transitions are implicit because the context is situated out of school, in the *ardinas*' practice, but elements of mathematics learned in school may be present in *ardinas'* associations, thereby linking the two domains. Because the process may be a reflexive one, for *ardinas* who are still in school this is a 'collateral transition' experience, in Beach's (1999) terms, although the changes the *ardinas* experience *within* the practice are what Beach calls 'encompassing transitions' because these occur within the boundaries of a social activity that is itself changing.

It is likely that the form-function shifts described by Saxe (1991) in his work with Brazilian candy sellers, are also present in the mathematical meanings constructed by the *ardinas* in their implicit collateral transitions between the two fields of practice, newspaper selling and school. Certainly their construals of their interactions with the interviewer (a woman, white, and Portuguese-speaking, like their teachers in school) suggest that the secondary Discourses of school mathematics and of the selling practice are very different. Santos and Matos write, 'We think that it is really difficult (for an outsider as in the case of an observer/researcher) to have deep access to the mathematical thinking process characteristic of the *ardinas*' practice, as the actual language-game of the practice does not include discussions about the mathematical facts we know to be present in that practice' (chapter 4). This difficulty seems to be present because of the lack of overt mathematical verbal communication in the borderland Discourse of the *ardinas*. They do not talk about the mathematics of their practice! That a borderland Discourse is present is indicated in their sharing of 'ways of acting, interacting, talking, valuing, and thinking, with associated objects, settings, and events, which are characteristic of people whose social practice it is' (Gee's 1992 definition of a Discourse, p. 91). And this Discourse is borderland because it is situated between the primary home Discourse and that of the school.

The tacit nature of the mathematics of the *ardinas'* selling practice makes this a difficult research field. In fact as Santos and Matos point out, when the interviewer is asking mathematical questions of an *ardina,* there are often two Discourses going on simultaneously. The *ardina* may try to approach the interviewer's questions about the mathematics of his practice from a school mathematical Discourse point of view (what Santos and Matos call the language-game of school mathematics). This is what Ntóni and Diku, as 'good informants' who are familiar with the school Discourse, did in several instances. But the mathematical thinking in these instances is clearly framed in a different way from that which may be tacitly inferred from the practice. Further, as illustrated in an episode with Ntoni, 'the *ardina* makes no mistake when he spontaneously performs calculations about the real value of an issue. But this does not happen when he has to explain his calculating process to the observer, at a time when he is still concerned with the sale and his attention is turned to this' (chapter 4).

The transitions involved in this collateral transition (Beach, 1999), learning the selling practice, appear to rely heavily on a borderland Discourse with old-timers, in which mathematical meanings are constructed from largely tacit aspects of actual participation in the practice. There is an implicit borderland apprenticeship system: there is a borderland Discourse without much explicit mathematical discourse. It is apparently up to an individual *ardina* how much of the school mathematical Discourse he will incorporate in his construction of mathematical meaning in his selling practice. However, because of its tacit nature, such links in meaning are not an overt part of the Discourse of selling.

The discrepancies in the Discourses of selling and of school mathematics are illustrated in the *ardinas'* dialogues about their profit of 20 escudos per newspaper in 1999. In the school Discourse, the selling price of 100 escudos, less the 80 escudos required by the newspaper agency, always results in a profit for the *ardina* of 20 escudos per newspaper: 100–80=20. In a dramatic way, the episodes in section 2.2.2 of chapter 4 illustrate that 100 minus 80 is *not* always equivalent to 20 in the Discourse of the selling practice. The social aspects of the two ways of obtaining the newspapers in the practice have very different meanings attached to them. Some of these meanings are positive and some are negative in each method, 'taking' or 'buying' newspapers from the agency. The result is that the *ardinas* themselves do not agree about which method is preferable. What is clear in the Discourse of the practice is that 100–80=20 as an abstract mathematical fact has little meaning for the *ardinas*: it does not help them to make the practical decisions that will determine what actually happens in their selling. Perhaps these striking dialogues more than any others in the book show *why* it is so often the case that in transition experiences, mathematical meanings appear to be subordinate to social ones.

5. SOME IMPLICATIONS FOR TRANSITIONING PROCESSES

At a deep level, Bishop (1991) has challenged the values and reasons commonly given for teaching the 'formal' mathematics curriculum. (His use of the term *formal* mathematics resonates with what I have elsewhere called *academic* mathematics; Presmeg, 1996). Bishop (1991) argued that it is the implicit nature of unrecognised mathematical values in mathematics education that causes difficulties for learners. The cases described in the empirical chapters of this book provide evidence that it is also values associated with various facets of the learning of mathematics that enable or constrain individuals in coming to grips with their transition processes between contexts for learning mathematics. It is the implicit nature of these values that causes some of the difficulties in these processes too. Kashif's mother did not know of her son's devaluation of her cultural capital, thus she was not helping him in his transition to schooling in England (chapter 5). Alberto's teacher was better able to understand his resistance to transition processes when she learned in a home visit of his successful participation in the family bakery in Mexico before the family emigrated to the USA (chapter 6), that is, when his rejection of the values associated with the process was illuminated.

'A new idea is meaningful to the extent that it makes connections with the individual's present knowledge', wrote Bishop (1985, p. 26). He proceeded to give examples of different kinds of possible connections, hinting at the different status afforded the teacher's knowledge in comparison with that of the student. However, the learner is the meaning maker, as Bishop pointed out (ibid.). The interplay of inner and outer experiences is confirmed in the meanings constructed by learners as described in this book. The importance of social dimensions of this meaning making is highlighted in the fact that an attempt to analyse the mathematical meanings of these learners in transition has yielded far more sociocultural connections than mathematical connections *per se*. Mathematical capital, while not devalued by the transitioners or their parents, was subordinate to cultural, symbolic, and linguistic capital as they tried to construct meanings for their transition experiences. The analysis suggests that increased awareness on the part of teachers and parents of the subtleties and difficulties, and of the roles values play, in meanings associated with these transitions, would help to facilitate learners' commitment to the processes involved. The analysis also suggests that an appreciation, by teachers, parents and learners alike, of the cultural capital that immigrant students and their parents *do* possess, would ameliorate some of the symbolic violence that immigrants often experience. Further, such an appreciation would aid learners in the construction of self images that are strong enough to undergird their shifts in meaning as they build new cultural capital in transition experiences. These issues are in stark relief in the cases of immigrant students, but the principles apply equally to learners in more subtle transitions experiences by minority learners or those whose cultural capital has been devalued in the wider society. In all these cases, mathematical meanings appear to be subordinate to sociocultural ones in the transition processes involved.

As Bishop pointed out (chapter 8), there are more productive ways to view cultural conflict than in terms of differences generating either-or choices. When this *exclusive* partitioning model is replaced by an *inclusive* partitioning that preserves heterogeneity, then cultural conflict may be viewed as a natural outcome of the opposition between parts of one harmonious whole. The complementarity of consonance and dissonance is seen in productive mediation through cultural interaction. A teacher embracing this model would encourage and facilitate cultural interaction in her classroom. She would view cultural diversity as a rich resource for learning mathematics, not, for instance, as an inconvenience to be ignored if possible, by placing immigrant students' desks out of her cone of vision and interaction (chapter 2).

In a similar way, *symbolic violence* experienced by students in transition between contexts when their cultural capital is devalued by significant others (as analysed in the present chapter) may be viewed as one pole of a continuum. The dissonance of this pole is balanced by the harmony of an opposite pole that could be termed *symbolic appreciation*. This pole is epitomized in the valorisation of cultural capital that students bring to the mathematics classroom. The process that links these two poles is one of *symbolic mediation*. It is a mediation because through transition processes that involve both structure and agency (chapter 8), teacher and students in a mathematics classroom can negotiate ways of facilitating the move from violence to appreciation of the individual cultural symbols that students bring with them. This dialectic process seems not only worthwhile, but essential in moving beyond the negative features of symbolic violence in classrooms. Some of the ways that these processes may be approached have been suggested in the theoretical chapters, and there are examples in the empirical chapters. For instance, one approach is through teacher visits to the homes of individual learners (chapter 6). As in all dialectic processes, this mediation is not a once-only occurrence, but an ongoing process of complementary dissonance and harmony being seen as one whole, a process that is continuously repeated. The theme of mediation is explored further by the three editors in the final chapter of this book.

REFERENCES

Abreu, G. de (1993). *The relationship between home and school mathematics in a farming community in rural Brazil.* Unpublished Ph.D. dissertation, University of Cambridge.

Adams, D.W. (1998). Education for extinction: American Indians and the boarding school experience, 1875–1928. *Harvard Educational Review, 68*(2), 263–264 (book notes).

Beach, K. (1999). Consequential transitions: A sociocultural expedition beyond transfer in education. *Review of Research in Education, 24,* 101–139.

Bishop, A.J. (1985). The social construction of meaning – A significant development for mathematics education? *For the Learning of Mathematics, 5*(1), 24–28.

Bishop, A.J. (1988). *Mathematical enculturation: A cultural perspective on mathematics education.* Dordrecht: Kluwer Academic Publishers.

Bishop, A.J. (1991). Mathematical values in the teaching process. In A.J. Bishop, S. Mellin-Olsen, & J. van Dormolen (Eds.), *Mathematical knowledge: Its growth through teaching*, pp. 195–214. Dordrecht: Kluwer Academic Publishers.

Bourdieu, P. (1995). *Language and symbolic power.* Cambridge, Massachusetts: Harvard University Press.

Civil, M. (1995). Connecting home and school: Funds of knowledge for mathematics teaching. In B. Denys and P. Laridon (Eds.), *Working group on cultural aspects in the learning of mathematics: Some current developments,* pp. 18–25. Monograph following the 19th Annual Meeting of the International Group for the Psychology of Mathematics Education, Recife, July 1995.

Clarke, D.J. (1998). Studying the classroom negotiation of meaning: Complementary accounts methodology. In A. Teppo (Ed.), *Qualitative research methods in mathematics education.* Reston, Virginia: National Council of Teachers of Mathematics, Monograph 9.

Cobb, P. & Yackel, E. (1995). Constructivist, emergent, and sociocultural perspectives in the context of developmental research. Plenary address, in D. Owens & G. Millsapps (Eds.), *Proceedings of the 17th Annual Meeting of the North American Chapter of the International Group for the Psychology of Mathematics Education*, Vol. 1, pp. 3–29.

Ellerton, N.F. & Clarkson, P.C. (1996). Language factors in mathematics teaching and learning. In A.J. Bishop et al. (Eds.), *International handbook of mathematics education,* pp. 987–1033. Dordrecht: Kluwer Academic Publishers.

Gee, J.P. (1992). *The social mind: Language, ideology, and social practice.* New York: Bergin & Garvey.

Grenfell, M., & James, D. (Eds.) (1998). *Bourdieu and education: Acts of practical theory.* London: Falmer Press.

Lerman, S. (1996). Intersubjectivity in mathematics learning: A challenge to the radical constructivist paradigm? *Journal for Research in Mathematics Education, 27*(2), 133–150.

Moschkovich, J. (1999). Supporting the participation of English language learners in mathematical discourse. *For the Learning of Mathematics, 19*(1), 11–19.

Nieto, S. (1996). *Affirming diversity. The sociopolitical context of multicultural education.* White Plains, New York: Longman.

Pimm, D. (1995). *Symbols and meanings in school mathematics.* London: Routledge.

Presmeg, N.C. (1988). School mathematics in culture-conflict situations. *Educational Studies in Mathematics, 19*(2), 163–177. Special edition.

Presmeg, N.C. (1996). *Ethnomathematics and academic mathematics: the didactic interface.* Paper presented in Working Group 21, The Teaching of Mathematics in Different Cultures, Subgroup 2, Preparing Teachers to Teach Diversity. Eighth International Congress on Mathematical Education, Seville, Spain, July 14–21, 1996.

Presmeg, N.C. (1998). Ethnomathematics in teacher education. *Journal of Mathematics Teacher Education, 1*(3), 317–339.

Reay, D. (1998). Cultural reproduction: Mothers' involvement in their children's primary schooling. In M. Grenfell & D. James (Eds.), *Bourdieu and education: Acts of practical theory.* London: Falmer Press.

Saxe, G.B. (1991). *Culture and cognitive development.* Hillsdale, New Jersey: Lawrence Erlbaum Associates.

Spindler, G.D. (1974). *Education and cultural process.* New York: Holt, Rinehart, and Wilson.

Valdés, G. (1998). The world outside and inside schools: Language and immigrant children. *Educational Researcher, 27*(6), 4–18.

Walkerdine, V. (1982). From context to text: A psychosemiotic approach to abstract thought. In M. Beveridge, (Ed.), *Children thinking through language,* pp. 129–155. London: Edward Arnold.

Walkerdine, V. (1988). *The mastery of reason: Cognitive development and the production of rationality.* London: Routledge.

Walkerdine, V. (1990). Difference, cognition, and mathematics education. *For the Learning of Mathematics, 10*(3), 51–56.

Walkerdine, V. (1997). Redefining the subject in situated cognition theory. In D. Kirshner & J.A. Whitson (Eds.), *Situated cognition: Social, semiotic, and psychological perspectives.* Hillsdale, New Jersey: Lawrence Erlbaum Associates.

EDITORS' POSTLUDE

THE SOCIOCULTURAL MEDIATION OF TRANSITION

1. TRANSITION REVISITED

In the first chapter of this book we discussed the reasons for choosing to build our work around the idea of transition. In brief we argued that

- this notion was useful to explore the impact that participation in multiple practices has on an individual's learning and uses of mathematics;
- the existing conceptions of transitions, such as Bronfenbrenner's, needed to be reframed in a more dynamic perspective.

In following a dynamic perspective we proposed

- that the way the person experiences a transition involves input from both individual and sociocultural structures;
- that both types of inputs need to be conceptualised as constantly evolving;
- that personal agency in the individual is paralleled by active social dynamics, such as inputs from significant others that are constantly negotiated during the participation in the practices.

Tentatively we suggested that Bronfenbrenner's notion of role transitions:

- if approached from a more dynamic perspective could be paralleled with Lave and Wenger's view of learning as 'legitimate peripheral participation' (Lave & Wenger, 1991), and
- if they occur within a practice they could not be understood without taking into account 'setting' transitions between practices.

The research described in this book focused predominantly on setting transitions, attempting to address issues related to the learner's movement between contexts of mathematical practices. In our view this was a major gap in sociocultural theories. The need to provide detailed situated accounts of the social and cultural structuring has generated research that focused attention on events within the practices. This in turn contributed to theoretical and methodological advances in interpretations of learning in terms of evolving participation within communities of practice. However, as stressed by Engeström (1999), 'people do not move only from the periphery to the center of a practice; they also and increasingly move between communities of practice' (p. 256). From a dynamic perspective it is likely that the construction of

G. de Abreu, A.J. Bishop, and N.C. Presmeg (eds.), Transitions Between Contexts of Mathematical Practices, 229–238.

participation in various practices is in fact a co-construction, or that there are interconnections that need to be understood. Unravelling what these connections might be and how they can be studied was our task in this book.

As we have seen in the five empirical chapters there are different aspects of transition that the authors have considered. However two main questions were addressed in all of them:

1. *What are the relevant features of the learner's experience of transition between mathematical practices?*

For example, we recall the metaphor of the 'invisible' shadow, where there seems to be a kind of silent negotiation that takes place between student and teacher. This is a phenomenon that is worthy of more exploration. Why do some students find themselves in this position? Perhaps they seek this situation while they adjust to language and cultural differences? It is well-known that second language learners need considerable time just listening to the new language use in the new context before wanting to contribute themselves. It is also a situation that involves various layers of prior knowledge and representations (mathematical, linguistic tools); classroom norms; social groups norms and assigned social identities (gender, ethnicity). Even if the language issues are not central, there will be new cultural orientations to adjust to and if one's 'original' cultural norms emphasise, for example, acquiescence and submissiveness, this would make it doubly difficult to engage experientially in the new context. As Oyserman and Markus remark (1998), transitions between contexts can bring changes in the social representations of personhood. Of particular relevance to our analysis, they stressed that meanings of competence change across contexts. Thus, in their view 'transition phases' are very often associated with feelings of loss of competence. The constructions of 'the other' are then central to the negotiation of participation in new contexts. They can contribute to inclusion (acceptance and negotiation of difference) or exclusion (silencing and repressing of the differences).

2. *What are the social dynamics and associated patterns of transition, involving for example, the role of significant others and social organisation?*

As we have seen from the empirical chapters, the roles of the parents, the teachers, and the peers are not as straightforward as those 'names' imply. They are not simply part of the 'social context' of the learner and thus apparently irrelevant to the immediate teacher/learner situation. They are each implicated in their roles as part of both agency and structure, and thus influence and shape interactively that situation. Parents influence both teachers and learners (see Abreu et. al.; Bishop; and Civil & Andrade chapters) in some contexts while in others the peers are more significant in influencing teachers and learners (see the examples in Santos & Matos; Gorgorió et al. and Bishop chapters).

In general the empirical chapters have illustrated the following key features of the transition experience:

- ***Knowledge historical transition*** – where the individuals or groups experienced a significant change in the mathematical practices due to historical changes within their own developing institutions or communities of practice, giving rise to cognitive conflicts and transitions. This notion for instance seemed important in circumstances where a homogeneous culture was shared, as was the case with the white-English parents supporting their children's school mathematical learning (Abreu et al., chapter 5) and with the *ardinas* in Cabo Verde, who moved from school to selling newspapers, and who also experienced changes in the selling practice when the newspaper organisation changed (Santos & Matos, chapter 4).
- ***Cultural transition*** – where the individuals or groups experienced a change in the cultural context, and thus other values, customs and practices became significant. This we can think of as the 'acculturation' process, with its accompanying cultural conflicts and transitions. This aspect of transition became more salient in situations where there is not one homogeneous culture. For instance, when the parents emigrated after being at school in their home country, the child might be exposed to one set of mathematical practices and representations at home and another set at school. Examples of these were given in the chapters by Abreu et al. (chapter 5), Bishop (chapter 3), Civil and Andrade (chapter 6) and Gorgorió et al. (chapter, 2).
- ***Social transition*** – where the individual or group interacted with different social groupings, and thus with different significant others, with the accompanying social conflicts and transitions. Here the emphasis was on social organisation and social roles. Examples of these were given by Bishop (chapter 3) and Santos and Matos (chapter 4), Civil and Andrade (chapter 6). As illustrated in Civil and Andrade's chapter the interactions that took place between women's groups in the mathematical workshops, for example, and the questions that the women raised were facilitated by the dyads and triads of women that came together to learn. They entered a number of social settings together, where they might not otherwise have ventured. Although there were marked differences in their knowledge and experience, they supported one another as learners in situations where once again, they might not otherwise have considered asking questions or engaging in the same manner. We saw this when men (spouses) were present – men became the 'learners' while women remained passive observers.
- ***Linguistic transition*** – where there was a significant shift in language structure, vocabulary and practice, resulting in language conflicts and transitions. Needless to say, language issues became more salient for learners instructed in a language different from their home one. However, monolingual speakers also had to move between everyday and mathematical discourse, such as in the case of some *ardinas* (chapter 4). At a discourse level it became clear that the borderland Discourse of the mathematics learning situation was the contested linguistic space of the transition experience.

It is likely that every learning transition involves all four aspects to some extent, but as we have seen, some are more important than others for any one individual in any one transition situation.

2. FROM TRANSITION TO MEDIATION

In the three theoretically focused chapters following the five empirical ones there are again different aspects of transition revisited and re-analysed by the authors. Each of the analyses raised important questions awaiting further analysis and research.

In chapter 7, for example, Abreu explored the link between current issues in cultural psychology and the kinds of issues that have emerged through our analysis of students' experiences of transition in mathematics learning. She outlined the theoretical trajectory from cross-cultural studies of mathematical cognition to studies of learning and use of mathematics in situated social practices to the studies of transitions between practices illustrated in this volume. Two particular questions highlighted were how a cultural tool mediates cognition and the need to take into account the role played by valorisation in this process. Drawing on examples from studies presented in the book she illustrated how the shift from a single practice to the transitions between practices, or transitions between contexts of a changing practice, shed light on the value-laden nature of mediation. In this case, Abreu argued, there is a strong mediating role played by social identities in the way the person learns and uses knowledge, which needs to be properly addressed in future research.

In chapter 8, Bishop analysed the idea of mathematical acculturation and discussed how the components of this construction have been modified as our research on transitions has progressed. Bishop discussed the theoretical developments concerning cultural conflict (from exclusive to inclusive partitioning), concerning acculturation (power as productive, and the teacher's role in creating and exercising this productive power) and concerning the borderland Discourse of the mathematics classroom 'workplace'. Finally he discussed some of the questions raised by considering the theoretical movement in education from one of intentional acculturation to one of cultural production.

In chapter 9, Presmeg discussed the shifts in meaning that take place during transition, through considering the shifts in meaning of immigrant learners, the role of parents in the construction of mathematical meaning, and through the role of language and associated symbolism. She concluded that from the research reported in the empirical chapters, mathematical meanings are subordinated to sociocultural ones in the transition processes. The question is then provoked as to whether this is the case in any learning situation, given that all learning involves transition.

However these three chapters have also revealed some common ideas – particularly relating to 'mediation', a construct of great importance to those who work in education. It is useful to look again at the four categories of transitions suggested by Beach (1999), and to examine the kinds of mediation involved in each.

In lateral transitions, individuals move between two related practices in a single direction. In this book, the cases of immigrant students, their parents, and their teachers all represent transitions of this kind. In chapters 2 and 3, Gorgorió et al. and Bishop respectively delineated the ways in which students constructed new identities in interaction with their teachers, their parents, and their peers. The social interaction was a powerful mediating factor in the construction and reconstruction of these new identities. The borderland Discourses of many of these interactions indicated to students new roles that were possible, thus pointing directions that these students might choose to take, such as 'shouting back' at peers so as to no longer feel like a 'shadow' in the classroom (Tra, chapter 3), or making the decision that she wants to be a mathematics teacher – but only of girls – in her new country (Saima, chapter 2), so that she will avoid the dilemma of different mathematics being suitable for girls and for boys. In mediating the cultural conflict of these transitions, parents and teachers of these students, too, need to see the processes as being part of one larger whole. This perception then, is also a mediating agent in the transitions of these parents and teachers. The transitions of parents were strikingly illustrated in the research of Abreu et al (chapter 5), and those of teachers in the home visits described by Civil et al. (chapter 6).

In collateral transitions, two or more related practices may be involved simultaneously, and the movement takes place repeatedly in both directions. Chapters 2, 3, 5, and 6 in this book contained many examples of such transitions, for instance where home mathematical practices differed from those at school and students were repeatedly moving between these contexts. The problematic nature of mediation in some of these transitions was made clear in the difficulties that parents experienced in helping their children with their homework in a way that resonated with the school practice of learning mathematics (chapter 5). An important factor in cases of successful mediation was parents' and teachers' awareness of the cultural conflicts between the home and school practices, and commitment to helping the student to bridge the gap, a process that might entail changes in both the home and the classroom practices. This process illustrates the mediation between symbolic violence and symbolic appreciation (chapter 9) that changes the dissonance of the former pole of this violence-appreciation continuum to the harmony of the latter one. Again, communication between parents and teachers seems to be a crucial element in bringing about these changes (chapter 6).

In encompassing transitions, changes in the practice itself and in the community of those involved in the practice entail new ways of operating, as in the changes over time in the *ardinas'* practice of selling newspapers. Two major elements that facilitated the mediation between different forms of practice were the *ardinas'* trust in an 'old-timer', one of their own community who understood both the old and the new ways of operating and the changes involved, and the establishment of supportive groups in the practice, for instance in the informal groups that developed as protection against 'pirates' when the boys returned their money and unsold newspapers to the agency in the new way of operating. Some of the implicit transitions between school mathematics and the selling practices may however be classified as collateral transitions because these transitions involve those ardinas who were still in school in a repeated movement between these two spheres.

Mediational transitions, in which an intentionally educative activity is designed to prepare participants for a future experience or activity, was a form of transition found only indirectly in the research described in this book. It is possible that mediational transitions of the form described by Civil and Andrade in the case of women's groups in their mathematical workshops could be used with groups of parents of mathematics learners in other settings, too, to facilitate the kinds of awareness of classroom mathematical practices that would aid parents in helping their children during lateral and collateral transitions.

In lateral and collateral transitions, changes in role and in setting are involved, in Bronfenbrenner's (1979) terms. In encompassing and mediational transitions, the predominant changes are in the roles of the participants in a largely unchanged setting. The analyses of transitions in this book indicate clearly that in these processes it is the whole person who is involved in constructing a new identity, with new meanings and values, finding a new niche in a new social role with new discourse patterns and power relationships. By theorising these transitions using various lenses, we have attempted to cast light on processes that may be implicit and unacknowledged even by the participants themselves. A major conclusion is that mathematical meanings of learners in transition are often subordinate to sociocultural meanings, depending on the linguistic capital, cultural capital, and symbolic capital of the learners.

Two particular meanings of mediation have been found to be significant in our analyses. Firstly we shall discuss **cultural mediation.**

The notion of cultural mediation is a core theme in Vygotskian formulations of learning. Vygotsky emphasised the cultural mediation of psychological functioning by claiming that 'higher mental functioning and human action in general are mediated by tools (or 'technical tools') and signs (or 'psychological tools') (Wertsch, 1991, p. 28).

Tools can be symbolic sign systems to represent mathematical ideas, such as counting systems or measuring systems, or they can be material artifacts such as calendars, calculators, and computers (Säljö, 1996). Vygotsky explained the mediating role of sign systems in human thinking, drawing an analogy with the role of physical tools which people create and use to interact with their environment. For him, a physical tool changes the way people interact with their physical environment, and indeed the structure of an activity, in the same way as a mental tool changes mental activity (Scribner & Cole, 1981; Vygotsky, 1978).

The second meaning of mediation emphasised in the various chapters is the **social**. Vygotsky also stressed the importance of social mediation, but he emphasised immediate social interactions (see Abreu, chapter 7). This notion has been associated with his view that learning takes place in a zone of proximal development, where performance is assisted by others who are more capable (Vygotsky, 1978). He viewed this interaction as the space where knowledge that existed in the social plane, was negotiated interpersonally, and gradually constructed in the intrapersonal plane.

But Vygotsky's original account of cultural and social mediation (or perhaps the way it was interpreted in Western research) needs expansion. As we have seen in the empirical chapters, cultural and social mediators are not used in value-free ways but relate to the valorisation and positioning of communities of practice in social orders (see also Abreu, chapter 7, and Abreu, 2000). In a certain way the sociological theory of Bourdieu (see Presmeg, chapter 9) provides an alternative to Vygotsky's formulation of mediation. The difference between this theory and Vygotsky's theory is that the tool is not dissociated from its social valorisation in Bourdieu's formulation. Thus, for Bourdieu the mediator is the '*habitus*', a notion that refers to *a set of dispositions which incline agents to act and react in certain ways* (Bourdieu, 1995, p. 12), and which will include both actual capital (cultural and linguistic tools) and symbolic capital (accumulated prestige or honour). Thus, the nature of the mediator is simultaneously cultural and social (in a wider sense that incorporates social orders). This is also in line with the cultural psychology approach Abreu outlined in chapter 7. She argued that studying the experience of transitions highlighted the value-laden nature of the cultural practices, and consequently, demonstrated the need to revise the dominant conception of mediation in Vygotsky-based theories.

It is clear (see chapter 9) that habitus, for an individual, is deep-seated and not amenable to easy change, since it involves a 'bodily nexus' and dispositions that become deeply ingrained in a person's identity through a lifetime. In this sense, habitus is the mediator between the individual and the society and culture in which it was formed: but it is not the mediator of change during transitions. Habitus may become the *vehicle* for such mediated change, although the process may be both lengthy and difficult, because it involves changing dispositions of which an individual may not even be aware.

Hence there is a need to consider the nature of the mediation processes through which habitus may be changed. In parallel with the construct of *cultural mediation*, we propose a process of *symbolic mediation* (chapter 9). The *symbolic violence* associated with the devaluation of an individual's cultural and linguistic capital is one pole of a continuum that may be considered to have *symbolic appreciation* as its opposite pole. This is a state of valorisation of an individual's cultural and linguistic capital: it may be associated with increased *symbolic capital* involving symbolic worth, and even honour and prestige. The perturbations involved in symbolic violence, in an individual's recognition that cultural capital has been in some way devalued and discredited, provide strong impetus for change. This is the beginning of symbolic mediation, a process of negotiation involving both agency and attempts to change structure, on the part of the individual, but including all those involved, such as peers and the teacher, and possibly parents. Bishop's chapter 3 abounds in examples of the start of this symbolic mediation, for instance when Tra 'yell(s) back and start(s) to annoy them' when her peers call her a 'nerd head'. She reports that she no longer feels like a 'shadow' in the class.

Abreu in chapter 7 introduces the concepts of identity chaining, competing identities and projected identities, which may contribute to explain possible impacts of processes of symbolic mediation in the learner. Thus, while practices that are structured around principles of symbolic appreciation could promote forms of identity chaining that are inclusive, its extreme counterpart of practices organised around principles of symbolic violence may promote forms of identity chaining that are exclusive. Furthermore, Abreu argues that the form through which this symbolic mediation operates in the context of the learner in development is projected identities. Social identities associated with cultural practices very often pre-exist in a social plane before being 'extended' to the newcomer or learner. Taking into account that the teacher is the principal actor in the context of classroom mathematical practices one can ask: What is the teacher's role in symbolic mediation? All the empirical chapters provide evidence that the teacher may be highly influential in this process.

When teachers learn more about the home cultural situations of their students (Civil & Andrade, chapter 6), not only may their entire view of particular students be changed, but they may become facilitators of this symbolic mediation process in their classrooms. It appears that in this way, teachers are at the heart of symbolic mediation in mathematics classrooms as in others. When teachers learn about their students' home cultures they change their views on who their students are. Thus, they are able to project different social identities on them. Alberto, the case study described by Civil and Andrade (chapter 6), after the home visits is not anymore just a student experiencing difficulties at school, in the eyes of his teacher he also became a competent player in outside school practices related to his home culture.

We have also seen in our studies the importance of the construction of social identities by learners. How does this relate to the idea of mediation? We have seen in chapter 8 some examples of the roles of teachers, parents and peers in mediating the transition process, and in facilitating the learners' construction of social identity. We have found that the learner's awareness of, selection of, and denotation of, the 'significant others' is crucial to their social identity development (see also Abreu et al. chapter 5, and Abreu, chapter 7). The 'significant other' is in some way the key feature of social mediation, and the enrichment and development of that idea within mathematics education is an important task for the future. This significant other however does not need to be physically present in the situation: the social other that one identifies can be evoked symbolically.

However, we have seen that the construct of 'significant other' reinforces not just the idea of a social 'other' but also that it is the learner who determines the significance. If we then return to the idea of *cultural mediation*, it can be argued that it is also the case that the significant cultural 'other' is not only recognised by the learner in transition but is also given significance by the learner. It is the learner who judges the significance for them of any aspects of the 'other' culture. (When we say 'cultural other' we are not referring to a particular person, although the cultural difference of significance may well be portrayed by people in their interactions with each other.)

So, although teachers, parents and similar cultural 'mediators' may judge certain cultural differences as being likely to be significant for the learners in transition, it is the learners themselves, through their observation of and their engagement in the different practices, who determine the 'significant cultural others'. These may or may not be associated with language or representational aspects, despite these being aspects that adults think are significant (see research on texts, diagrams, etc…), but they may be more behaviour-related, value-related, or relationship related. As we can see in the three vignettes with which we began this book, Severina, Saima and Mohamed are busy choosing, and giving significance to, the (for them) significant aspects of the transitional cultural space.

3. TRANSITION, MEDIATION AND NEW RESEARCH CHALLENGES

It is customary following a theoretical analysis on mathematics education, and particularly one that is concerned to develop new theoretical ideas, to conclude with some implications for practice.

It might have seemed tempting therefore, in this final section, to have developed some grand over-arching sets of pedagogical principles applicable to every mathematics teaching situation. That however would have flown in the face of all we have seen discussed in the earlier chapters, all of which have emphasised the situated nature of transition and the locally contested, and interactive, nature of mediation.

We believe that our contributions to knowledge in the mathematics education field from this book, as a whole, and from our collaborative project, are principally theoretical and research-focused. This of course does not of itself rule out any implications for practice. Indeed our search was not specifically for novel practice procedures, although each of the five empirically-focussed chapters do shed light on some more, and some less, appropriate mediation practices. Rather we believe that our collaborative endeavour has enabled us to reveal some exciting new avenues for theoretical and research exploration, discussed in the three theoretically focussed chapters.

By focusing on learners in what we can think of as non-standard transition situations, we have been able to reveal features of mathematics learning which are normally not in the foreground. However this development has come at a price, and that price concerns complexities and dilemmas involved in doing this kind of research. Sociocultural mediation as interactive, emphasises the learner's unique contribution to the mediation process, as well as to the transition experience. One consequence of this is that there may be little value in research that though addressing learning in sociocultural contexts neglects the learner's experience of transition and the implicated mediating processes. All the empirical chapters refer in some way to the extreme complexity of undertaking transition research in the sociocultural domain. Furthermore it has been made clear to us throughout this project that research itself is always formulated and prosecuted in a contested political space, particularly in transition learning situations.

Nevertheless this research has been done, and as a result some new theoretical formulations have been achieved. These in their turn will challenge and change research approaches, foci, and sites. Once again it is not feasible for us to present an overarching set of research principles applicable to every transition situation. But we can certainly throw our weight behind developments that urge awareness of social, cultural, political and ethical dilemmas in research on mathematics teaching and learning.

We would argue that rather than considering a mathematics learning research study/project in isolation from its social, cultural, and political context, it is crucially important to recognise that context as part of the mediated transitional space. It is therefore equally important in any such research project to be always assessing the role and significance of those who create, structure and negotiate that space. For, as Presmeg writes, 'Mathematical capital, while not devalued by the transitioners or their parents, was subordinate to cultural, symbolic and linguistic capital as they tried to construct meanings for their transition experiences' (chapter 9).

Whether that idea generalises to every mathematics learning situation will have to be tested by others but we have no doubt that as students such as Severina, Saima and Mohamed have shown us, it holds true for our specific studies on transition in mathematics learning.

REFERENCES

Abreu, G. de (2000). Relationships between macro and micro sociocultural contexts: implications for the study of interactions in the mathematics classroom. *Educational Studies in Mathematics, 41*(1), 1–29.

Beach, King (1999). Consequential transitions: a sociocultural expedition beyond transfer in education. *Review of Research in Education, 24,* 101–139.

Bourdieu, P. (1995). *Language and symbolic power.* Cambridge, Massachusetts: Harvard University Press.

Bronfenbrenner, U. (1979). *The ecology of human development.* Cambridge, Mass.: Harvard University Press.

Engeström, Y. (1999). Situated learning at the threshold of the new millennium. In J. Bliss, R. Säljö, & P. Light (Eds.), *Learning sites: social and technological resources for learning* (pp. 249–257). Oxford: Elsevier Science.

Lave, J., & Wenger, E. (1991). *Situated learning and legitimate peripheral participation.* Cambridge: Cambridge University Press.

Oyserman, D., & Markus, H.R. (1998). Self as social representation. In U. Flick (Ed.), *The psychology of the social* (pp. 107–125). Cambridge: Cambridge University Press.

Säljö, R. (1996). *Learning and discourse: a sociocultural perspective*: The British Psychological Society Education Section.

Scribner, S., & Cole, M. (1981). *The psychology of literacy.* Cambridge, Mass: Harvard University Press.

Vygotsky, L. (1978). *Mind in society: the development of higher psychological processes.* Cambridge, Mass: Harvard University Press.

Wertsch, J.V. (1991). *Voices of the mind.* Cambridge: Cambridge University Press.

AUTHOR INDEX

SUBJECT INDEX

Mathematics Education Library

Managing Editor: A.J. Bishop, Melbourne, Australia

1. H. Freudenthal: *Didactical Phenomenology of Mathematical Structures.* 1983
   ISBN 90-277-1535-1; Pb 90-277-2261-7
2. B. Christiansen, A. G. Howson and M. Otte (eds.): *Perspectives on Mathematics Education.* Papers submitted by Members of the Bacomet Group. 1986.
   ISBN 90-277-1929-2; Pb 90-277-2118-1
3. A. Treffers: *Three Dimensions.* A Model of Goal and Theory Description in Mathematics Instruction The Wiskobas Project. 1987 ISBN 90-277-2165-3
4. S. Mellin-Olsen: *The Politics of Mathematics Education.* 1987
   ISBN 90-277-2350-8
5. E. Fischbein: *Intuition in Science and Mathematics.* An Educational Approach. 1987
   ISBN 90-277-2506-3
6. A.J. Bishop: *Mathematical Enculturation.* A Cultural Perspective on Mathematics Education. 1988 ISBN 90-277-2646-9; Pb (1991) 0-7923-1270-8
7. E. von Glasersfeld (ed.): *Radical Constructivism in Mathematics Education.* 1991
   ISBN 0-7923-1257-0
8. L. Streefland: *Fractions in Realistic Mathematics Education.* A Paradigm of Developmental Research. 1991 ISBN 0-7923-1282-1
9. H. Freudenthal: *Revisiting Mathematics Education.* China Lectures. 1991
   ISBN 0-7923-1299-6
10. A.J. Bishop, S. Mellin-Olsen and J. van Dormolen (eds.): *Mathematical Knowledge: Its Growth Through Teaching.* 1991 ISBN 0-7923-1344-5
11. D. Tall (ed.): *Advanced Mathematical Thinking.* 1991 ISBN 0-7923-1456-5
12. R. Kapadia and M. Borovcnik (eds.): *Chance Encounters: Probability in Education.* 1991 ISBN 0-7923-1474-3
13. R. Biehler, R.W. Scholz, R. Sträßer and B. Winkelmann (eds.): *Didactics of Mathematics as a Scientific Discipline.* 1994 ISBN 0-7923-2613-X
14. S. Lerman (ed.): *Cultural Perspectives on the Mathematics Classroom.* 1994
    ISBN 0-7923-2931-7
15. O. Skovsmose: *Towards a Philosophy of Critical Mathematics Education.* 1994
    ISBN 0-7923-2932-5
16. H. Mansfield, N.A. Pateman and N. Bednarz (eds.): *Mathematics for Tomorrow's Young Children.* International Perspectives on Curriculum. 1996
    ISBN 0-7923-3998-3
17. R. Noss and C. Hoyles: *Windows on Mathematical Meanings.* Learning Cultures and Computers. 1996 ISBN 0-7923-4073-6; Pb 0-7923-4074-4

18. N. Bednarz, C. Kieran and L. Lee (eds.): *Approaches to Algebra.* Perspectives for Research and Teaching. 1996 ISBN 0-7923-4145-7; Pb ISBN 0-7923-4168-6

19. G. Brousseau: *Theory of Didactical Situations in Mathematics.* Didactique des Mathématiques 19701990. Edited and translated by N. Balacheff, M. Cooper, R. Sutherland and V. Warfield. 1997 ISBN 0-7923-4526-6

20. T. Brown: *Mathematics Education and Language.* Interpreting Hermeneutics and Post-Structuralism. 1997 ISBN 0-7923-4554-1
Second Revised Edition. 2001 Pb ISBN 0-7923-6969-6

21. D. Coben, J. O'Donoghue and G.E. FitzSimons (eds.): *Perspectives on Adults Learning Mathematics.* Research and Practice. 2000 ISBN 0-7923-6415-5

22. R. Sutherland, T. Rojano, A. Bell and R. Lins (eds.): *Perspectives on School Algebra.* 2000 ISBN 0-7923-6462-7

23. J.-L. Dorier (ed.): *On the Teaching of Linear Algebra.* 2000 ISBN 0-7923-6539-9

24. A. Bessot and J. Ridgway (eds.): *Education for Mathematics in the Workplace.* 2000 ISBN 0-7923-6663-8

25. D. Clarke (ed.): *Perspectives on Practice and Meaning in Mathematics and Science Classrooms.* 2001 ISBN 0-7923-6938-6; Pb ISBN 0-7923-6939-4

26. J. Adler: *Teaching Mathematics in Multilingual Classrooms.* 2001 ISBN 0-7923-7079-1; Pb ISBN 0-7923-7080-5

27. G. de Abreu, A.J. Bishop and N.C. Presmeg (eds.): *Transitions Between Contexts of Mathematical Practices.* 2001 ISBN 0-7923-7185-2

KLUWER ACADEMIC PUBLISHERS – DORDRECHT / BOSTON / LONDON